Table of Contents

Forward 1
Introduction 3
It's My Story and I'm Sticking To It 7
I Do Believe In Ghosts. I Do, I Do… 15
   Ghosts and Residual Hauntings 20
   Jefferson, Texas – Spooksville, USA 26
   Active Hauntings 32
   Messenger Ghosts 44
   Helper Ghosts 53
   Vortexes & Ley Lines 59
   Portals 65
   Stagnant Energy 68
   Ghost Hunting 101 70
   Finding Ghosts Intuitively 79
Baby, I'm Stuck On You 91
   What Are Attached Entities 94
   Entity Motives: The Truth About Entities 101
   Cords: How Entities Attach & Control 104
   Why Entities Attach 111
   Incubus and Succubus 130
   Entities And Inanimate Objects 136
   Alien Entities 140
   Demons 144
   Entities and Mental Disorders 147
   Dissociative Disorders 156
What To Do, What To Do… I Don't Need All Of This
   Excitement! 163
   The Use Of Intention In Healing 168
   Clearing Stagnant Energy From a Location 170
   Protecting Your Environment 175

- Working With Ghosts 180
- Detecting And Eliminating An Entity From Your Auric Field 193
- Helping Someone Get Rid Of An Attached Entity For Good 201
- That's All, Folks 211
  - Home Protection Kits 215
  - About Dr. Rita Louise PhD, ND 219
  - Other Works by Dr. Rita Louise 223

DARK ANGELS
An Insider's Guide To Ghosts, Spirits and Attached Entities

Rita Louise, PhD

Copyright © 2009 Rita Louise, Ph.D.
All rights reserved.

Editing by Wayne Laliberté, MS, MBA
Book Design by Mitchel Whitington
Cover Art modified by Justin Baker

All rights reserved. No part of this book may be reproduced or transmitted in any form or by any means, electronic or mechanical, including photocopying, recording, or by any information storage and retrieval system, without permission in writing from the copyright owner.

ISBN-10 # 0-9758649-8-X
ISBN-13 # 978-0-9758649-8-2
Library of Congress Control Number: 2009908562

Third Printing March 2013

Second Printing April 2010
Printed in USA

Published by SoulHealer Press
Dallas, TX
SoulHealer Press
www.soulhealer.com

To all of the people, with or without bodies, who have made this book come to life.

~ Vive La Liberté ~

Forward

Many contemporary "blockbuster" authors achieved their success by creating fictional worlds where the natural and the supernatural collide – witness the works of J.K. Rowling and her *Harry Potter* series, Charlaine Harris' *Sookie Stackhouse* books, and Stephenie Meyer's world of *Twilight*. The public seems fascinated by the thought of the paranormal intruding into our normal, everyday life.

There is an old adage stating, "Truth is stranger than fiction," however, and it is certainly true of our world. We are surrounded by supernatural beings, whether we choose to acknowledge them or not. Entities such as ghosts, angels, demons and other spiritual forces exist among us, influence our lives, and interact with us. Our world abounds with ethereal beings such as these – reality can rival any fictional tale.

Too often our thoughts on the supernatural are based on Hollywood movies or television shows that have been crafted only to entertain, not to instruct. It can be hard to determine exactly where the facts end and a fanciful story begins.

Thankfully, we have Dr. Rita. In her latest book, *Dark Angels*, she pulls back the veil and reveals the darker side of the spiritual realm. Rather than simply discuss the attributes of these forces, she uses her extensive knowledge and experience to take us on a journey through the world of the unseen – to explore, to interact, but most importantly… to understand.

– Mitchel Whitington
Author of *A Ghost in My Suitcase*

Introduction

Books about angels and spirit guides abound. These books discuss who they are, how they interact with us, and how we can communicate with them. Angels and spirit guides, nevertheless, are not the only kids on the block. *Dark Angels* addresses the bad kids in town. Who are these bad kids? Well, they are the ones who live on the dark side; namely, ghosts, attached entities, and demons. Like the former, ghosts, entities and demons also interact with us and as you read on, you will discover the ways and means of their appearance in your life.

Whether we are talking about angels, guides, ghosts or demons, they all share one thing in common. They are all entities. An entity is a being without a corporeal (physical) body. Entities come in all shapes and sizes. They can look just like you and me and appear to have come from planet Earth or they can appear quite alien and look like something right out of a science fiction movie.

All entities exist outside our 3-dimensional universe and outside the world of matter and form. It has only been in recent years that quantum physics has validated the concept of "subtle energies" within our universe that are unseen. It is also theorized that everything in our universe is made of different frequencies of vibrating energy. Items which exist within our physical world are believed to vibrate at a low frequency and appear solid to us. In turn, our thoughts, feelings and emotions are said to vibrate at a much higher frequency. We are able to sense them and experience them, but at this time, we are unable to measure them.

Conversely, entities exist in the world of subtle energy. Their presence in our world is difficult if not impossible to measure using today's technology. But this does not mean they do not exist. Like our thoughts, feelings and emotions, they are something we are able to sense and experience in our lives.

The varieties of entities are classified based upon their energetic vibration. The higher the vibration, the more open and expansive the energy seems. The lower the vibration, the energy is less expansive and will seem contracted or dense in comparison. If an entity is vibrating at a high vibratory level we experience them as filled with love, light, peace, wisdom and universal truths. If their vibration is low they are often seen as being consumed with ego, pride, greed, and insecurity or are out seeking power and control.

These non-corporeal beings can be categorized by identifying their vibratory rate. In their order of hierarchy they include:

        1) Angels
        2) Spirit guides
        3) Ghosts
        4) Attached entities
        5) Demons

Introduction

There are no hard and fast lines to separate one kind of entity from another. Instead, there are a wide range of vibrations overlapping each category. For example, when talking about socio-economic groups, one family may be from the upper middle class and another from the lower. Both, however, are classified as being part of the middle class, even though their income and way of life may differ greatly.

Angels and spirit guides are thought to present a high positive vibratory rate. Conversely, attached entities and demons have a low negative rate. This leaves ghosts somewhere in the middle. Some are helpful, kind and beneficial, and others can be mean, cruel or spiteful.

Entities exist outside of the physical world; it is impossible to view them with the naked eye. The concept of proving an entity's existence resides on the fringe among ghost hunters, psychics and others who have experienced them in their lives. Unfortunately for each of these groups, technology has hindered their ability to document an entity's existence through the use of systematic measurements. Photographs, audio recordings and videotaping of encounters with entities is all that current technology can offer. Wholehearted individuals who endeavor to prove their existence using these forms of documentation are often discounted or their proof is regarded as a hoax.

Despite tangible evidence, there are hundreds, if not thousands, of individuals who claim to have encountered an entity at one time or another. It may be a miraculous circumstance, a bump in the night, or some kind of unexplainable behavior. Each individual is describing an interaction with a being without a body. These individuals are talking about their encounter with an entity.

Dark Angels

Before we move on and delve into the world of Dark Angels, let me say one last thing. What you are going to read in the following pages might seem a bit fanciful. To some, the concepts and stories being told may seem outlandish or just way too "out there" to be true. Even as I write these words and share my stories with you, they sometimes seem too bizarre even for me to believe. But the truth is, all of the tales I will be telling are based on my experiences.

It's My Story and I'm Sticking To It

Dark Angels

> *"When I'm asked 'Are you afraid of the dark?' my answer is 'No, I'm afraid of what is in the dark."*
> – Barry Fitzgerald

You may ask how I know so much about ghosts, entities and the other odd and assorted beings without bodies. And the fact is, I have been surrounded by them my entire life. I was two or three years old when, unbeknownst to me, I experienced my first encounter. While I would love to tell you all about it right now, we will come back to that story when we discuss attached entities later in the book.

The first encounter I had when I was two or three was not the only one I have ever experienced. At the age of 10, my family moved into our first real home. It was an old Victorian house and was built in 1898. The house was massive and was made up of four levels. It started with a full basement. The main floor comprised the kitchen, living and dining rooms. The second floor held four bedrooms and then the attic contained another three full bedrooms. Creaky and drafty as any house built during that time tended to be, the one thing my seven siblings and I were all quite aware of was the "dead guy" in the basement.

Growing up, everyone hated going down to the basement. Unfortunately, it was where the extra refrigerator and chest freezer were kept. If we happened to be the unfortunate soul who inadvertently walked into the kitchen when my mother needed something such as a bottle of soda or a head of lettuce from the extra fridge down stairs, the unlucky new arrival to the kitchen was volunteered for the task.

Young, old, male or female, our method of navigating the basement was always the same.

First, we would open the basement door as wide as possible. This added extra light to the basement stairs and helped further illuminate the route we would be taking. This tactic also meant that you did not have to mess around with turning a doorknob on your way back up the stairs, thus assisting us in a rapid escape. It also reduced the risk of the dreaded possibility the door would accidentally close behind you on your descent. We definitely did not want anything to interfere with our rapid escape from that nasty place!

Before entering the stairwell, we would illuminate the overhead light. This was done by reaching through the doorway with one hand while remaining safely in the confines of the kitchen. Then it was time to head down the stairs. We would run down the first few stairs as fast as possible, turn the corner, and then bound down the remaining steps. At the same time, we would raise one hand over our head, as if by instinct, and would pull the cord to the ceiling light that hung over the utility sink to the right.

Once we were safely down the stairs, we made a quick dash across the room. In route, we would reach up to the left and flick the switch, which turned on the light for the basement proper. Still going at full speed we would do whatever we needed to do to get out of there as fast as possible, turning off each light as we ascended from Hades.

Sometimes however, in our haste to get in and out of the creep-filled environment, we would grab the wrong thing. A bottle of soda instead of a dozen eggs was a mistake we would deeply regret. If this horrid error did happen, we would have to go "way down there" again and correct our mistake. This would also seem to be the times when somehow the upstairs door would inexplicably close and lock behind us.

I was not the only one who found myself trapped in the basement stairwell feverishly trying to get out. I recall many times in which the door would mysteriously lock behind me. Upon my ascent I would begin frantically pounding on the door and pulling on the knob all to no avail. In those intensively fear filled moments, thoughts of the dead guy "getting me" would run rampant through my mind. Then after a few minutes of unrestrained terror, I would hear devilish laughter coming from the other side of the door. Oh yeah…but that part was not from a ghost. God, you have to love your family.

Of the ten family members who lived in the house, the only one who did not sense the Dead Guy's energy or was not bothered by going into the basement was my mother. She would spend hours down there doing laundry. We never understood how she could manage it, especially since the Dead Guy could always be found right next to where she hung the clothes to dry. I still cringe when I think about the Dead Guy.

Many, many years later at a family get-together, I happened to make a side comment about the Dead Guy. This was never talked about when we all lived in the house. To my surprise, everyone knew exactly who I was talking about and where his energy resided in the basement.

My interaction with entities did not end there although at times I wish it did. Several years later, I started college at Oswego State University at the tender age of seventeen. Oswego State is located in upstate New York. It was during my second year of school that I encountered entities again – except this time, it made the Dead Guy seem like a long lost friend. Sophomores were allowed to move off campus and find housing of our own. It seemed like the opportunity of a lifetime to this freshly minted sophomore.

Oswego is a small college town located 30 miles north of Syracuse, New York. It is literally right on the shores of Lake Ontario. Due to its proximity to the lake it was subject to severe "lake effect." This meant if a storm rolled in during the winter it would pick up moisture from the lake and dump it on the town. Thirteen feet of snow per year was considered normal in Oswego.

By a stroke of enormous luck, I found an available room in a house with three other students. From the outside, it was like a dream come true. The apartment was located in the upstairs portion of an old Victorian house. It was fully furnished and situated right on the main street. Without a car to drive, the location was ideal. It was only five blocks from the downtown area and best of all, the corner in front of our house was a stop on the bus route, which went all the way to the college campus. This meant you could stand inside and wait for the bus to arrive instead of standing outside and freezing in the cold winter air. In addition to its great location, the rent could not be beat – or could it?

The apartment sat on top of a business that sold what I interpreted as "mos-o-leum" (spelling is modified to make a point). I thought it was some kind of linoleum – like the kind you would put down on your kitchen floor. It was not until my pronunciation was amusingly corrected by a friend who informed me that they did not sell "mos-o-leum" they sold "mausoleums." I could have passed out...they sold crypts to hold dead people.

Learning that simple fact was chilling enough! Not long after acquiring that tidbit of knowledge, I was out with some friends and I began a conversation with someone who was raised in the town. He asked me where I lived and I told him, on the corner of 5th and Bridge. Looking me right in the eyes with a deadpan expression on his face, he said, "Oh, you live in the old Funeral Parlor." My eyes became wide as my heart sank. A large part of me did not want to believe what he had to say, but deep down I knew it was true. I will share some tales about my stay in Oswego later in the book when we talk about residual energy ghosts.

My next encounter with entities was during my studies at the Berkeley Psychic Institute (BPI) in California. This time my experience with them was on a much more conscious/seeing/knowing level. At BPI I learned to control, understand and validate the information I was receiving about a human client. During sessions with my clients I began to see entities around some of them. What I discovered in working with these individuals was the entities were somehow attached to them. Even back then it was easy for me to see how these "attached entities" interacted with my client and how these entities affected their lives.

During my time at BPI I also volunteered to be part of a house healing team. It was our responsibility to go into a physical location, evaluate it from an energetic point of view, and then heal or clear any anomalies or disturbar detected. We would often encounter ghosts, vorte⋎ entities in people's homes.

In my private practice as a naturopath, medical intuitive and clairvoyant as well as in my work with Metroplex Paranormal Investigations (a ghost hunting team in the Dallas area), I have visited many locations with ghosts hanging around as well as individuals who have entities attached to them. Over the past few years, I have encountered or have become even more acutely aware of entities, than ever before. This is especially true in my private practice. The number of people who have come to me with entities attached to them has definitely grown. I am constantly astounded by the tremendous impact these entities have on my client's well-being.

Before I move on and begin talking about hauntings, and ghosts in particular, I want to share one final piece of information. In the stories I tell as I work as a member of a ghost hunting investigation team, let it be known that I am referring to the "Metroplex Paranormal Investigations" (aka, "Metroplex) where I work as their psychic investigator. Our protocol requires me to go into a location "cold."

When investigating a location "cold", I am taken to a location or given a physical address. That is it. I am not briefed nor informed of any of the reported activity in the location. This approach keeps my views, insights and responses uncontaminated by stories or rumors told by the homeowner or other investigators. If after investigating what I perceive matches the homeowner's concerns, then the homeowner, the group and I can acknowledge that there must be something going on at the location even if nothing appears using traditional ghost hunting methods.

I Do Believe In Ghosts.
I Do, I Do…

Dark Angels

I Do Believe In Ghosts. I Do, I Do...

"He's stuck, that's what it is. He's in between worlds. You know it happens sometimes that the spirit gets yanked out so fast that the essence still feels it has work to do here"
— Oda Mae Brown

The best-known form of a "being without a body" is a ghost. Ghosts are believed to be the spirit or soul of a person who has remained on Earth after death. Ghost stories and stories of haunted locations have been told for generations. Modern media has increased our knowledge and understanding of ghosts by providing us with a large number of examples of ghostly hauntings – some are believable but most are not.

I remember as a child there was a television show called "The Sixth Sense" starring Gary Collins as Dr. Michael Rhodes. Dr. Rhodes was a college professor who investigated mysteries involving extra-sensory perception (ESP), spirits, possessions and other paranormal activities.

During a typical episode there would, of course, be a murder and Dr. Rhodes would be called in to help. Entering a crime scene he would often see the likeness of the murder victim – their ghostly image suspended in mid-air. The ghost would provide him with a key piece of information about the murder or a clue to the perpetrator of the crime. I was led to believe the wispy, see-thro˙ image of the deceased floating around a room as pr˙ in the show IS how a ghost would make its appe˙

Dark Angels

Whether it was in one of the haunted houses I lived in, or an investigation into the strange disturbances reported by a homeowner, I have yet to see what is often depicted by the media. Ghosts, like all entities, live in the world of subtle energy, which to the casual observer exists outside our physical world and beyond the range of our five senses. When we see a ghost, we are not seeing them with our physical eyes. Instead we are using our alternative eyes – our clairvoyance – to view them.

I am often reminded of the x-ray vision glasses that were popular in the '60's and 70's when I think of seeing ghosts. When you put them on, it was claimed you could see through people's clothes, your skin, just about anything. When we see a ghost clairvoyantly, it is like we have put our ghost glasses on, intentionally or accidentally and then view the world through the lenses of the glasses. Looking through them makes the invisible visible even if for only a brief moment.

Ghosts can and do interact with our physical world. They are the things that can go bump in the night. Actually, they go bump at any time of the day. Ghosts have been known to open doors, hide items (only to have them reappear later on), move furniture, and turn lights on and off. It is claimed by many that their voices can be heard on tape and their images caught on film. When utilizing traditional ghost hunting equipment, such as cameras and tape recorders, nothing is visually detected in the room. It is not until the pictures are viewed, or the recordings analyzed, that the presence of a ghost is revealed.

Like alien abductions and the whole UFO phenomena, most ghostly encounters and their documentation is systematically debunked and dismissed by mainstream science as fake. Despite this, there are countless testimonies from credible witnesses around the world attesting to their personal encounters with ghosts.

As for myself, I believe ghosts and entities are real. I have had too many profound experiences with them for them to be only a figment of my imagination. For the non-believer and skeptic at heart, as you learn more about ghosts in general and how you too can perhaps catch one on film or detect one intuitively, you might decide to change your mind and become a believer like me.

Ghosts and Residual Hauntings

When a ghost has taken residence in a location it is referred to as a haunting. All hauntings fall into two distinct categories; residual and active. Although they may seem similar to the untrained observer, they are very different in their nature.

A residual haunting is the most common type of ghostly event. Although frequently thought of as a haunting by a discarnate soul (the soul of someone who has passed) this is not the case. A residual haunting is in essence the energetic imprint of a person or event on the environment that can be seen, felt and perhaps heard.

Ghosts that traditionally make up a residual haunting are tied to a site, a room or even a piece of furniture. There are not any actual ghosts, spirits or entities involved in this type of haunting. What is being encountered in these situations is a recording of a past event as opposed to interacting with a being without a body. Vicki Issacs, the founder of Metroplex Paranormal Investigations, describes a residual haunting as being similar to a loop of film which keeps playing the same footage over and over.

While it is unclear how imprinted energy is recorded in an environment, there are a number of theories as to why. Pete Haviland, the founder of Lone Star Spirits, a ghost-hunting group based in Houston, Texas, believes a residual haunting is trapped energy that is anchored to a location. This can include the recording of moments of intense emotions, such as battles like the one in Gettysburg. It could document the energy of a murder or other violent or traumatic event. It could even be the imprint of pleasurable emotional energy, such a big fun filled party or intense sexual energy. Imprints can also be recorded when an activity is consistently repeated such as sitting in a specific "favorite" chair, walking up a flight of stairs or cooking in a kitchen.

Imprinted energy can also be attached to the personal effects of an individual, such as a ring or watch, or even a piece of furniture. Many people have seen mentalists or psychics on television ask for an individual's ring or watch and will proceed to provide them with intimate information about themselves or their lives. What they are doing is tapping into the residual or imprinted energy of the person being held within the object.

Interaction with this type of ghost is limited. Like a holographic movie playing in the middle of a room, they are unaware of your presence. Residual hauntings can include not only a visual impression of the person or event, but also its associated sounds, smells and emotions. Dress and demeanor can readily describe these individuals. Data such as who they are, why they are there, or what happened to them is often easy to surmise. While direct conversations with ghosts that make up a residual haunting is impossible, it is possible to tap into the remnants of energy they left behind and amass a great deal of information about them.

The presence of a residual ghost can literally raise the hair on the back of your neck. Since they are residual in nature they pose no threat to anyone who may encounter them.

The Dead Guy – Revisited

Remember the Dead Guy, the ghost who lived in the basement of the house where I grew up? Our experiences with him were a classic example of a residual haunting. Why was he there? Was there a reason his energy was trapped in the basement of our home? Well, long after we moved into the house and I was aware of the Dead Guy, this is what I heard...

Mr. Gruburg, the man who ran the deli around the corner, told me someone had been murdered in our house. Being young and remarkably gullible I was scared half to death. Was it true? Had someone died in our house or was it just a tale he told to scare a young impressionable kid? I'll never know. But if someone had been killed in our home, it was most certainly the Dead Guy!

I never told anyone in my family about what Mr. Gruburg shared with me that day. His comments did have one dramatic effect on me. They increased my level of fear and trepidation about going down into the now even more haunted basement. Even though I never told anyone about the murder and intrigue that happened in our house, on some levels they all knew.

The Old Funeral Parlor – Revisited

By the end of my stay in the old funeral parlor in

I Do Believe In Ghosts. I Do, I Do...

Oswego, NY my roommates and I had determined we were being visited by at least three separate entities. Looking back at my time there, I can now see the reality of what we were experiencing. I can tell you with certainty the ghosts were residual in nature. Residual or not they scared the crap out of me, my roommates and anyone else who visited the house.

As you learned earlier, the old funeral parlor was housed in an old Victorian home, with a business on the first floor and our apartment on the upper level. Entering into the heavy front door of the house, one would immediately see two inner doors. One led to the "floor covering (mosoleum)" business downstairs; while the other went to our unit.

If you opened the door leading to our apartment, you were greeted by a steep set of carpeted stairs that took you to the upstairs landing and our living area. When I moved into the apartment I decided to hang a string of small bells on the entry doorknob of our unit. Thus, if someone happened to open our door we would be notified of their entry well in advance.

Oh my God... There were countless days and nights where my roommates and I, and even our guests, would hear what sounded like the exterior front door opening. We could hear the sound of air as it would rush in from the outside, creating a kind of suction between the door. This suction would cause the bells to rattle in a particular way. The sound was loud enough that it always caught our attention. Then we would stop and listen to see if we had an unexpected guest. Perhaps it was someone from the business downstairs entering the main door.

Unfortunately for us, as we sat in silent anticipation of the new arrival, we would not hear a door at all, ours or the one downstairs. Instead, what we did hear next was the creaking of wood as if someone, or something, was making its way up the stairs to our unit. For some of our friends it was more than enough for them to never step foot in our apartment again. For us, we would nervously just wait and pray for the sound to stop.

For some reason, toward the end of the school year, my roommates and I got the bright idea to borrow a Ouija board. We figured with only a few weeks left to finish school, and our anticipated departure (or should I say escape) from the house, we decided it would be fun to try and contact the spirits who resided in the apartment.

After several unsuccessful attempts to contact the undead that lived in the apartment, my roommate (Val and I) decided to go to bed. We each disappeared into our respective rooms. Lying in bed, trying to mind my own business and not think about the ghosts, I could hear the quiet creaking of what sounded like footsteps down the hallway. As it neared my room, the creaking was joined by what sounded like someone tapping its knuckles against the wall. Curled up in bed with the covers over my head, I called out, "Val, come here," to which she responded, "No, you come here." Neither of us said another word to each other that night.

Many years later when I evaluated this situation intuitively, the experience we had in the hallway of our apartment was a residual haunting. The energy belonged to an older man who had been the janitor (or maintenance man) for the funeral parlor. Day after day, he would labor up those stairs and down the long narrow hallway leading to our kitchen. (Our kitchen had in the past been the utility room for the funeral parlor complete with an oversized sink.) I could also see in my mind's eyes the image of him carrying his cleaning bucket filled with dirty water. As he walked down the hall the bucket would bump into the walls. This residual sound imprint left us with the impression that he was tapping his fingers against the wall.

Jefferson, Texas – Spooksville, USA

Now if you are looking for ghosts, you might want to consider going to Jefferson, the most haunted city in Texas. While I want to provide you with a flavor of this ghost filled city, I do not want to spoil your fun in discovering the wide range of entities you may encounter while there. To me, knowing all about a location's history, especially knowing what kinds of activities have been reported takes away from your ability to discover its true history for yourself.

The Grove

The most spook-filled location per square footage of the many haunted sites in Jefferson has to be The Grove. The Grove is owned by Mitchel and Tami Whitington and was built in 1861. The property has an extended history dating back long before the house was ever built. It also has an unusual number of regular visitors who date back just as far in time.

Upon my first visit to the Grove I encountered no less than fifteen different ghostly anomalies. Many of these were residual. For example, when I first arrived and got out of my car I looked up at the front porch and was greeted by one of the former (and deceased) property owners. He was dressed in brown pants and a white shirt with suspenders. He also maintained a white moustache with a full beard. He was standing in front of the front door and watched me and several members of Metroplex as we got out of the car and started to unload our gear.

I did not want to be seen as rude, so I waved to him as I made my way up the walkway to the front door. Later I learned this man is often seen standing on the porch, or just inside the front door, where he keeps watch over the house and the neighborhood.

In the dining room I saw a woman. She appeared to be in her mid-twenties wearing a long blue skirt, white blouse and an apron. Her dress gave me the impression that she had lived in the house around the turn of the century. She kept walking into the room and would tend an invisible fire in the fireplace.

In the yard, there was the presence of an older woman. I call her "Grandma." She could be intuitively sensed along with the energy of a young girl who liked to visit with her. These are just a couple of the ghosts who live there. This list does not include the Native Americans, the secret lover and many others presences found within the bounds of their property.

The Jefferson Hotel

Down the street from the Grove is the Jefferson Hotel. Without giving away too much information about this site I'll just tell you it is also filled with ghosts. One of my personal favorites is a ghost I refer to as Bruno. Bruno is a big man when looked at intuitively. He is dressed in dark colored pants and a white shirt with his sleeves rolled up past his elbows. Bruno appeared to be at least 6'4" tall and has muscles bulging everywhere. If he were alive today he could easily be a professional bodybuilder.

I met Bruno on my first visit to the hotel. I was there with Metroplex. We were being filmed as part of a documentary about the many ghosts that inhabit Jefferson. Carefully walking around the first floor of the hotel and moving in and out of a number of the downstairs rooms, I discovered some had ghosts; others seemed at the time to be empty. It was when I headed upstairs to the second floor where I met Bruno.

Bruno was standing about five steps from the top of the stairwell. He stood right in the middle of the stairs. His arms were crossed. My first impression was that he was controlling the movement of people up and down the stairs. And based upon his size and demeanor, I'll bet if he said no, he meant business. If you get a chance to interact with him ask him why is he there? I think you will be quite surprised. I know I was.

I have introduced all kinds of people since my first meeting with him. These people were believers and nonbelievers alike. If you ever have the opportunity to visit Jefferson, try to stay at the Jefferson Hotel. It is worth your time. And if you do visit, do not forget to say hi to Bruno.

Beauty And The Book

During another investigation in Jefferson with Metroplex we stopped inside a small shop called Beauty And The Book. The concept of the store was interesting and oddly functional. Half the shop was a bookstore while the other half was a beauty salon. Upon meeting the owners, I wondered if their uniqueness would be more interesting than anything I might find in the store. Thankfully, they did have something that sparked my interest.

At one point as I made my way around the shop one of the investigators called me to the front of the store. Mark was standing by the large plate glass window that overlooked the street outside. I walked over to where Mark indicated and could feel the presence of an old woman in front of the window.

In my mind's eye I could see her rocking back and forth very slowly in a rocking chair. Her clothes seemed too large for her frail body and appeared old, tattered and mismatched. I got the impression she did not leave the house very often. Instead spent her time looking out the large window watching the world as it went by. If I were to characterize what her energy felt like to me into something probably very politically incorrect, she felt like a "cat lady." It was easy for me to envision her preferring the company of her many cats while detesting any interaction with people.

I walked slowly through the small space in front of the window. At one point I stopped abruptly, looked at Mark and said, "she's a thrower." A bit confused by my comment, Mark wanted me to clarify what I meant. I told him I received the impression that in response to what she saw through the window she would react in anger and would throw things. I added that it would not surprise me if things "flew" off the shelves and onto the floor. From what I understand, according to the owner, they frequently did exactly that.

Beauty And The Book has now moved to another location, and the shop has changed hands. While there may be new owners, I am sure the ghost of the old woman will still be rocking in her rocking chair in front of the large plate glass window.

The Haunted Bed

Simple ordinary objects can also carry the energy of their owners. During a multiple day investigation, I stayed at the House Of The Seasons. This is a bed and breakfast also located in Jefferson, Texas. Embarrassing as this may be to admit and contradictory as it may sound... I am a chicken when it comes to all things related to ghosts. As the group's psychic I was "fortunate" enough to sleep in the room that was believed to have the most activity. I was overwhelmed with excitement when I was "volunteered" to sleep in what had been described as the "haunted bed." Personally, I would have been my pleasure to pass this task to another member of the investigation team. Being a professional, I agreed on one condition. Another member of the group had to stay in the room with me just in case.

Sleep was hard for me to achieve during the few nights we were there. I could not tell if it was because there was something odd going on in the room, which kept me awake, or if it was because of the haunted bed. The one constant was that every time I was in the bed, my legs felt very heavy and cramped. When I looked at them intuitively, I could see braces, like the kind worn by Forest Gump as a child in the movie of the same name. I dismissed my observations as being a bit "out there" and tried to sleep while at the same time wait for something spooky to happen.

At the end of our visit, it was time for us to share our findings with the property owner. As for the room itself, we had nothing exciting to report. There was the residual energy of a woman and two children, in the room adjacent to ours, but we did not detect anything abnormal in the room we occupied.

The owner then inquired about the bed. Fishing for something to say, I described the stiffness I experienced in my legs and the sense of having braces on them.

To my surprise, I learned the bed had been the property of Colonel Benjamin H. Epperson. He was the original owner of the house. Colonel Epperson suffered from polio. This disease caused extreme pain and weakness in his legs. Many individuals with polio also suffer from sleep disorders, and based upon my inability to sleep soundly, he may have suffered from this condition. In addition, and this is the weird part, I discovered the Colonel spent many years of his life donning leg braces to support the weakened muscles in his legs.

Active Hauntings

Active hauntings make up less than 20 percent of all ghostly encounters. They are rare, especially when compared to the number of residual hauntings one might encounter. Unlike a residual haunting where what is being seen, felt or even heard is a recording or imprint of energy on an item or the environment...an active haunting involves real spirits or entities. During a residual haunting, any ghost detected is always unaware of your presence. This is not the case during an active haunting. Here the ghosts are aware of you and are known to interact with the living. These entities are true ghosts!

A true Ghost or earthbound spirit is an entity who, for whatever reason, is trapped between worlds. They are caught somewhere between their life here on earth and the spiritual planes inhabited by angels, spirit guides and those who have moved into the light. Individuals stuck between worlds are often emotionally attached to a person, place or thing within this world. They can find themselves bound to the earth for a number of reasons. The most common reasons include:

- They may not know they are dead.
- They mistakenly are waiting for a loved one to pass.
- They stay to take care of some unfinished business or for a life issue to be resolved.
- They feel guilty about leaving too soon or leaving family members behind without care.
- They are afraid to take their next step and move into the light; i.e., they are afraid of the unknown.
- They do not want to be judged for past events.
- They do not want to let go of what they know and what is familiar to them.
- They fear losing control.
- They are unwilling to surrender, thus letting go of their egos.

Individuals who encounter ghosts may also find themselves being visited by two other kinds of spirits; messenger ghosts and helper ghosts. These "ghosts" are not stuck between the physical world and the hereafter. Messenger ghosts are the spirit of the recently deceased who are saying goodbye to their friends and family. Helper ghosts are spirits who have transcended the physical plane, gone into the light and have returned.

In the world of ghost hunting it is believed that ghosts need energy to manifest, to move objects or speak. This energy can come from an electrical current, low grade radiation or even from strong emotional energy such as excitement or fear. Surprisingly, ghosts also have the ability to drain energy from a power source, such as batteries. During an investigation into a haunted location it is not uncommon to have a set of brand new batteries mysteriously lose their charge.

Regardless of the type of ghost involved, it is common during an active haunting for items to be moved. Stories of lights or electrical appliances being turn on and off, hearing voices or sounds out of nowhere, being touched, tickled, stroked or shoved are often told. I have heard reports of keys or other personal items disappearing only to turn up later on. Pictures of the departed loved one may mysteriously fall over or the scent of a specific perfume or tobacco may be noticed. Some people describe the feeling of an invisible "someone" standing next to them in a room or sitting on their bed. Depending on the individual ghost, sometimes they make their presence very clearly known and at other times they may remain quietly in the background.

Turning lights on and off, opening doors or creating disturbances in an electrical appliance are actions a ghost will take to get your attention. The do not mean to scare you. Most ghosts are not evil despite their stereotypical media fame. Surprisingly, they reflect the personality of the individual even after their death. If they were fun filled and liked to tell bad jokes in life they will act the same way in the afterworld. If they were mean, cruel or insensitive in life they can reflect this in the way in which they interact with the people they encounter in the environment.

Ghosts that make up an active haunting are often associated with the people who live in a troubled location. It might be a deceased family member or a close friend. You might wonder if it is your mother or father. It might be a grandparent or a friend, who recently passed, and is stopping by to pay you a visit. And yes, any of these could be the case.

A ghost can come to us for a number of other reasons. For example, in the movie *Poltergeist*, the ghost was drawn to the love and light that Carol Ann possessed. Sometimes an earthbound spirit will interact with an individual because it is looking for help. The ghost can detect the sensitivity of the person to its presence and it hopes to communicate its wants and needs.

Active ghosts can stay in close proximity to specific individual for days, weeks, months and even years after their passing. A recently deceased individual might have his or her presence felt among friends and family a good 30 days after passing. Sometimes a ghost will stop by for a quick visit just to say hello. Others may make their presence known for a brief period of time. For example, if an individual is going through a rough phase in life he or she may desire the love, support and guidance of their late mother. And in answer to their prayers she comes to lend a hand.

Some ghosts however, are known to associate with an individual for extended periods if not for their entire lives. These ghosts can be thought of as being "attached" to the individual. In some instances there may have been a past life relationship between the ghost and the affected individual. In these situations a bond or "contract" was created in the earlier lifetime and has endured.

Children are also very vulnerable to having a ghostly entity attach. Does your child have an invisible playmate? From the perspective of ghostly presences do you really think your child is making it up? Children are especially sensitive to the presence of ghosts. Sometimes children open doorways to the other side and end up getting more than they wanted. Monsters in the closet or under the bed could be an indication of a ghost in your child's room. Whether the ghost is naughty or nice, its presence can leave your child feeling scared and apprehensive.

In the next section we will explore a number of other reasons why a ghost may choose to attach itself to an individual. In terms of an active haunting, however, if a ghostly entity is "attached" to an individual, it is common to hear complaints about occurrences in other previous residences. These individuals are not constantly moving into haunted houses. They are taking their own ghosts with them from place to place.

Sheppard

One of the most memorable investigations I have ever participated in with Metroplex was at the residence of a young couple named Sheppard and Marissa. They had recently moved into a home together. It was not long after they took up residence "together" that problems began.

I was unaware of what had been transpiring in the house and from the outside it seemed to be a nice home in a congenial neighborhood. Things changed quickly as soon as I walked through the front door. My stomach immediately tightened and I began to feel nauseous. Strange as this may sound, this is always a great indicator that something juicy is happening. Continuing into the living room and then into the family room of their home the air around me felt thick and heavy.

I do not recall when it was, but it was not long into our investigation when I noticed the distinct presence of a man. He was a very angry man. This man seemed older, like someone who could have been a father, uncle or older friend to the couple. Whoever he was, his energy was filled with rage. His anger, violence and abuse saturated the home. In life, it seemed as if he ruled by intimidation and used fear to keep people in line. Apparently he was doing the same after his demise.

Walking from room to room I would sense his energy in one area of a room watching our investigative team. If anyone in the group drew close to him, he would literally disappear only to reappear in yet another room. We followed this very active ghost around the house. We were like the paparazzi using my intuition and the other group members tracking him with their meters, cameras and video equipment. We were fascinated with every move he made and did not want to miss capturing him on film or detecting him with our meters. In a bizarre way his actions seemed humorous to me. But my sense of humor was something he did not seem to appreciate at all.

At one point during our exploration, Kristi and Melissa, two other members of the investigative team, walked into a small bathroom on the second floor of the home. They seemed to feel his presence in the room and quickly called me in. Instinctively we stood facing each other in a small circle in the middle of the bathroom with the ghost in the center of the circle.

His presence was undeniable. Kristi, Melissa and I talked about how strong his energy was and how effortless it was to feel. I think we stood around him for about a minute. Then he disappeared without warning. Kristi and Melissa could easily tell that he had left. At the same time I heard a small chime ring in my head. It reminded me of the kind of sound you would expect to hear if someone was waving a magic wand. This sound was a clear signal of the sudden departure of the very active ghost from the room.

Connected to the other side of the bathroom was the master bedroom. Walking in and making my way around to the far side of the bed I could again feel his presence. This time instead of standing next to him as I had done in the past, I stepped into his energy field. I allowed his presence, his energy, to saturate me.

Looking through "his" eyes I saw the image of a woman whom he was beating severely. I could feel the anger, hatred and rage he had for this woman. The next thing I knew, I reached out into the air and grabbed her invisible shirt. I began pounding my fist into the unseen face of this defenseless woman. I felt like I was under a spell or as if someone else were moving my body.

I took a step back and shook off whatever had possessed me. Looking down my hands were red, hot and swollen. I have never had something like that happen to me before this investigation and have not had it occur again in any subsequent ones. It was an extremely intense experience.

Throughout the investigation, Sheppard could be seen drinking beer. This point was something the entire investigative team noticed. He drank the whole time we were there. He seemed visibly upset by what was going on and stated the beer was helping to settle his nerves. As the evening progressed, he became quite intoxicated. We did not think too much of it at the time, but its importance became clear during a second investigation to this home.

The investigation went by quickly and a follow-up investigation was scheduled for Sheppard and Marissa's home. I was unable to attend but another psychic, a good friend named Robert was there. It was reported to me that much of what Robert detected as he walked through the home was consistent with my findings. During the second investigation Robert also stepped into the ghost's energy field. He then called out the entity's name. It was Ron.

To everyone's amazement, especially Sheppard's, Ron was his father's name. Sheppard revealed that his father was an alcoholic and a very abusive man in life. It was also divulged that the problems they experienced seemed to intensify whenever Sheppard drank.

A few months after our investigation the couple split up. Sheppard moved out leaving Marissa in the home alone. She was afraid at first but it was not long before Marissa noticed something she had not expected. Within a few short days all of the strange occurrences in the house came to an abrupt stop. Marissa was relieved.

Pam's Grandmother

It was a warm Saturday evening. I was part of an investigation evaluating a very impressive new home owned by Pam and Mike. Pam and Mike seemed like a very nice, well-grounded couple who would not be swayed by their imagination. Things however, just were not right in their home. They and their three children experienced odd happenings in the house that they could not explain.

I observed several areas, which contained residual energy as I made my way from room to room. First, I sensed a boy of about ten in the upstairs hallway of the house. When I walked into the master bedroom, I detected the presence of another young child. This one was about five or six years old. My impression was this young boy would often climb into bed with Pam and Mike. At the time, this did not seem odd to me. The owners had a young son who I assumed climbed into their bed at night.

The family room was located on the second floor of the house. It was furnished with a sofa, a couple of chairs and a lot of exercise equipment. To the left of the room was a wall that was lined with bookshelves. On the top of the shelves was a series of antique pictures. I assumed the pictures captured the images of past family members. Based upon the style of photograph it appeared they were taken sometime in the early 1900's.

One particular picture, in the middle of the group, caught my eye. It had a very strange feeling to it. Whenever I looked at it my skin would shiver.

At the conclusion of the investigation, I provided the homeowners with a summary of what I detected. I told them about a teenage boy I observed in the upstairs hallway and the creepy antique picture in the family room. The couple then asked about the young boy they had experienced in their bedroom. I thought it was one of their children and was surprised to discover this was not the case.

I began to delve further into the energy of the child. I had the distinct impression that he was somehow associated with Pam in particular. I asked Pam if she had had a miscarriage or lost a child at a young age. Lost children often make their presence known to family members, especially their mothers. That was not the case for Pam. I was bewildered that a direct connection to her had to be ruled out.

I probed into the child's energy even deeper. The image of the eerie woman in the picture upstairs kept coming into my mind. I began to follow the path that was unfolding before me and evaluated the relationship between the small child and the lady in the creepy picture.

It was easy for me to discern that the child was connected to the woman in the picture. It was also clear to me that he was not her child. On the surface, this did not make any sense. The emotional connection to the woman was obvious and distinct. I could feel the love and safety the child felt when around her. He seemed scared, lonely and vulnerable and looked to her for love, reassurance and emotional support. I could readily sense why he had such a strong connection to her.

I asked Pam if she knew anything about the woman in the picture. It was her grandmother. I asked if her grandmother was in a position where she interacted with small children on a regular basis. Perhaps a position where a relationship could form between her and the child. To our surprise her grandmother had been a Sunday school teacher for many years.

We also learned that Pam was a teacher as well. She taught elementary school. Pam exhibited the same kind of loving support for the children in her care as her grandmother. I concluded that the child was attracted to Pam because she reminded him of his cherished friend, Pam's grandmother. As he did in the past, the child was now coming to Pam for the love and solace he needed and desired.

The Boogie Man

It was a balmy spring morning when I received a call from a young mom named Shannon. Shannon lived in Bermuda and was at the end of her wits. She was very concerned about her three-year-old daughter Jessica. Jessica complained of a scary man who would come into her room at night – a virtual boogieman. Jessica was frightened so badly she would start crying and screaming when she was put into her bed at night.

Shannon was afraid that something was very wrong with her daughter and was desperately looking for assistance. She tried everything she could to help calm Jessica's fears. Nothing seemed to work. Shannon was tired and extremely frustrated when she decided to contact me.

In the beginning of my investigation, I told Shannon that I believed her child was very open to sensing the energy of spirits. I could also sense that Jessica had already made the acquaintance of some "invisible friends." This fact was confirmed by Shannon.

I explained to Shannon that these spiritual beings did not bother or scare her child. Instead, I believed she enjoyed the time she spent with them. Shannon had accepted her daughter's little friends since their first appearance. She also made it a point not to discourage her from interacting with them. What Shannon did not realize is that when children open doors into the spirit world there are times, when uninvited souls come through. This definitely was the present issue for her daughter.

In addition to her invisible friends, I informed Shannon that Jessica was also being visited by a man who I would have to say was in his middle to late forties. The man came in when Jessica opened the door and let her little playmates come through. He came right in. Right into her bedroom.

I evaluated the man's energy and he did not seem mean or violent. I also did not sense that he would harm her in any way. I felt like he wanted her to see him, to acknowledge his presence and perhaps wanted to interact with her from time to time.

I suggested to Shannon that she might talk to her daughter about the man. I believed that if she could get her daughter to begin talking about the man, in a short amount of time, they could create some kind of relationship with him. They needed to question the man. They could ask him about what he looked like and maybe even ask him his name. This could ease Jessica's fear and anxiety. Instead of being a stranger, the man would now be a known entity. Over time, they both could interact with him.

Apparently, Shannon took my advice. Within a few

days, I received a jubilant call from her. Her daughter slept peacefully in her own bed by herself without trouble for the first time in over a year. I am so happy Shannon took the risk to talk to her daughter. Now both are having good restful nights sleeping.

Messenger Ghosts

Another group of ghostly entities are what I like to refer to as messenger ghosts. These are the spirits of people who have recently passed away and have come back to say goodbye to their loved ones. It is common to hear stories of individuals who claim to feel, see and even interact with someone who is recently departed. Messenger ghosts do travel into the light and transition fully into their next existence once the business they came back to do is complete. It is believed this process can take up to 30 days to conclude.

For some ghosts their chance to say goodbye to their friends and family occurs while their loved ones are sleeping – in their dream state. This is often the easiest way in which communication between the living and the dead can be made. Some people recount the sensation of a warm loving presence around them. Others speak of or the feeling of someone sitting down next to them on the bed. A number of people report seeing their loved one standing in a doorway or walking down a hall. They may hear the favorite song of the recently departed on the radio. This may be another way in which they say goodbye. Sometimes a personal item once thought lost will make a miraculous appearance.

It is hard to say if everyone associated with a Messenger Ghost is visited during this transitional period. Those who do recall a visitation are often more sensitive (more open) to having this kind of experience. They may also be less embarrassed to reveal what has happened to them. The evidence for this is still unclear.

Uncle Johnny

I am not a person who remembers my dreams or can easily recall dreaming at all. One night however, I had a very strange dream. I dreamt of my Uncle Johnny who I had not seen in years. He was coming to see me to have me do a Reiki healing session on him. Reiki is a form of hands on healing which is often used to support physical, mental and emotional healing. I remember wondering in my dream if my uncle even knew of Reiki. I was also perplexed how he knew I did this kind of work in the first place.

I can still recall watching him as he approached. It was odd. The setting of the dream was not anywhere in particular. The space around him was an empty void and was dark grey in color. It seemed as if he was surrounded by some kind of mist or fog.

As my dream progressed, he had started out some distance from me and kept moving toward me until he was about ten feet away. Then for some reason he stopped. He told me he did not need the healing session anymore. Then without any further conversation, he turned and walked away.

I immediately awoke from the dream and looked at the clock. It was 5:30 in the morning. The psychic part of my mind knew he had passed away or was in the process of passing. The cognitive part of me tried to rationalize what I had experienced. I decided not to call anyone to have them confirm or deny my suspicion.

Around mid-morning, my phone rang. It was my mother calling. She wanted to let me know my Uncle Johnny had just passed. She was just as surprised as I was. What neither of us knew was he had been diagnosed with cancer six weeks earlier. It was some time during the night of my dream, or very early in the morning, when he took his last breath.

Even when you do intuitive work like I do, it can be very hard to accept all of the messages we receive. I knew as soon as I woke from the dream that my uncle was in the process of dying. But it was the phone call that verified my dream and suspicion.

Healing and energy work can help us let go of emotional energy that can keep us from moving forward. In the case of someone who is in the last stage of life, it can help them to let go of their fears and support their transition. I guess by the time my uncle reached me he was able to address his fears and move on.

I do feel blessed that I was able be with him in his last moments and to be able to say goodbye to him as he passed.

Larry

It was December 30; just two days after a big snowstorm blanketed the entire Pacific Northwest. I found myself sitting quietly in front of my computer catching up on the never-ending pile of work that lay waiting for me.

I Do Believe In Ghosts. I Do, I Do…

In the midst of my work, I heard a loud noise coming from the street. I looked up from my computer and saw a fire truck driving down the quiet little dead end street and into the cul-de-sac at the end. At the time I didn't think anything unusual was in play. With the snow, flooding and downed power lines, I assumed that someone on the cul-de-sac had some kind of problem. Although my curiosity and intuition were aroused, I dismissed it.

It was when two fire department medic trucks drove down the street that I knew there was much more going on than just a little disturbance. Getting up from my work, I asked my husband if he knew what was happening outside. He did not. Uncharacteristically I left the house, walked down the street and headed toward the cul-de-sac at the end.

When I reached the corner and saw the fire truck and the medical units parked in the middle of the snow covered circle of the cul-de-sac, I felt the need to visit Larry and Janie's house to make sure that everything was all right. I turned back down the street, and toward their yard. Heading to their back door, my typical entrance, I found myself overwhelmed with an urge to get there as fast as possible. I began running across their yard through a foot of virgin snow.

I approached the back door of my neighbor's house. The door swung open and my neighbor Janie, in her robe, stepped into the doorway. Her eyes were red and swollen. Her grandfather, to whom she was very close, lived with her. When I saw the look on her face, the analytical part of my brain took over saying something was the matter with Grandpa. As I drew nearer to the door, Janie uttered two words… "It's Larry." Running in and closing the door behind me, I gave Janie a giant hug while not knowing, or even fathoming, what was still to be learned.

Looking up after the long embrace, I could see, in the middle of Janie's living room, six paramedics kneeling around Larry. One gentleman was pumping air into Larry's mouth and lungs, while another massaged and pumped his chest. One of the men was holding bags of saline solution that dripped slowly into Larry's veins. Another man regulated the air supply and cleaned out Larry's mouth and throat of fluids, while another operated the medical supplies, handing the required medications to the paramedics. The final gentleman, the one obviously in charge of the situation, knelt over an EKG machine where he checked and monitored Larry's vital signs. He also manned the defibrillator, which is the machine that shocks the heart, with the hope that a heartbeat would be restarted when a jolt of an electrical current is run through the body.

After surveying the scene, I stood with Janie, my mouth open in shock. I was unable to believe what I was seeing with my own two eyes. This man, who had visited my home a couple of days before, was alive and healthy, was now immobile on the floor. I felt helpless.

It was perplexing, because it took me a good 5 minutes before I was able to remember that I was a healer and that I could try to help my friend by working on him energetically.

With my intuition engaged, I again surveyed the room. The first thing I noticed was that the paramedics were well-grounded as they worked. Janie, Grandpa and I, were in a state of shock. I noticed that the room was filled with the energy of non-permission – non-permission for us to assist the paramedics in any way. I saw it was this energy that kept me from going right into healer mode from the very first moment. Once I recognized this energy, I decided I was not in agreement with it. I grounded myself and called back my power. This allowed me to acknowledge the energy and internally begin moving past it.

I said hello to each of the men that were working on my friend telepathically. I wanted to step closer, to put my hands on his lifeless body, but the paramedics had created an energetic barrier around their workspace. I respected their need for separation and proceeded to work on Larry from where I stood.

I continued observing the situation with my body being a little bit more grounded and my psychic glasses on. I was astonished by what I saw. To begin, when I again looked over to where Larry's body was immobile on the floor, I saw Larry's spirit floating up and over his physical body. It hovered about 6-12 inches over his body and had a soft golden color to it which sparkled with electricity. Although it did not look like my friend in the physical sense of the word, the energy did have a long body-like shape to it.

My first reaction to seeing this was to try to put Larry back into his physical body. This is something that I will normally do during a healing session. So the idea of trying to put Larry back into his body did not seem far out to me at all. I tried pushing on his essence from across the room. I tried to push him back into his body, but he would not budge. I tried to give him a grounding cord, the energetic connection that connects us as spirit to the physical world, and then tried pulling him back into his body by pulling on the cord. That did not work either.

Failing success with those two tactics, I was becoming a bit frantic. I began doubting my own skills and abilities. I wanted Larry to come back to life, take a deep breath, and sit up on the living room floor. I started talking to Larry, telling him that Janie was really upset and that we really were not in agreement to him passing right at that moment. I continued trying to push him back into his body, telling him to get back in with each push.

I can laugh at it now, but as I continued trying to push him back in his body, I started arguing with him. When I looked to where I heard his voice, it was then that I saw Larry standing behind me in the kitchen. He was in a spot that I had seen him stand many times before. His left hand was resting on the kitchen counter. His back was toward the door. I watched him as he stood there, a look of determination and an ornery expression on his face saying, "G__ D___ it, I don't want to go back!"

It was interesting because I stopped. I stopped trying to push him back into his body. I stopped trying to convince him that he needed to come back. In life, I knew that once he had made up his mind, I would not be able to change it, so why would it be any different in death. I let him go.

When I looked back towards his body on the floor in the living room, I noticed that it had shifted. Instead of his energy being in the long and humanoid shape just above his physical body, it had become more rounded, and a hole had formed in the center. I had never experienced this before and stood watching as this energy, his energy, started slowly moving out and across the room. I could feel the intensity of his life force energy growing larger and larger as the diameter of the circle increased.

As I watched his energy growing and expanding, it reminded me of a stone being dropped into still water, vibrational rings forming, growing in ever-larger concentric circles around the place where the stone was initially dropped. I never thought of our own life force energy being dissipated back into the universe this way.

I also felt a heaviness. There was a denseness to the energy as it headed toward Janie and myself. I knew the denseness we were experiencing was Larry. I watched as his golden electric like energy continued to expand through the room. I waited for it to engulf us in the place we stood.

When his energy had expanded far enough to reach where we stood in the kitchen, I started feeling faint. I wanted to share what I was experiencing with Janie. But I felt that it was not the right time or the place to start with my hocus-pocus stuff by telling her that we were standing in the middle of Larry's energy and that he was saying good-bye.

I have never before stood in this type of energy, which makes it a hard to describe. It was a very different experience from being in a room filled with guides, angels or even departed family members. The energy was much denser. It had a very distinct electrical charge to it. I received the impression that this is the way we transform our energy from the denseness of our 3-dimensional existence and bring it back into its multi-dimensionality. It is as if we were going from a state of compression to a state of freedom. I felt blessed to have had the opportunity to feel the expansion of his energy.

Larry left the house. When his energy had passed by us, and literally through us, I knew in my heart of hearts that it was over. The paramedics worked on him unceasingly for about 45 minutes. They should be commended for the effort that they expended trying to bring him back to life. I had never been in such close physical proximity to someone going through the death process. It was a very intense experience and it was one that I will always remember. It is not often that you are afforded the opportunity to be with someone as they pass. Thank you Larry for sharing this with me.

Helper Ghosts

Helper ghosts are the final group of ghostly presences we will be discussing. Helper ghosts are the spirits of individuals who have taken their next step and gone into the light. They have shed the negativity, ego and other limiting characteristics of their former life that may have kept a true ghost trapped here on earth.

Helper ghosts return to the earth (in nonphysical form) in order to help or support individuals. The nature of a helper ghost is similar to that of a spirit guide with one exception. Their vibration is not as highly evolved as what is seen when looking at the vibration of a spirit guide.

Helper ghosts are often related to the person or family member they are visiting. It is not uncommon to have a parent, grandparent, aunt, uncle, brother, sister or child pay us a visit. Close friends of the family are also known to make appearances to people who have been important to them in their former life.

Many times these ghosts will appear when there are important things going on in our lives like weddings or the birth of a child. Sometimes they can be found watching over us. I have regularly experienced deceased parents or grandparents watching over their children, grandchildren or great-grandchildren. This is especially true when the child is an infant. At other times they come to say hello and want to let everyone know they are fine.

Jim's Dad

Several years ago I was in the process of moving into a newly purchased home with my now ex-husband Jim.

One evening as I was stood in the kitchen unpacking dishes I detected the presence of a man sitting at the kitchen table. When I looked toward where I felt this man's presence I got the distinct impression it was Jim's father.

I never met Jim's dad. He passed away a good 10 years earlier. This was long before Jim and I met and married. Although I was not 100% certain of what I was sensing, the man sitting at the table reminded me so much of my husband I could not imagine it being anyone else.

Calling through the house I asked, "Honey, if your Dad were in the house would he be sitting at the kitchen table drinking a cup of coffee?" The response I received was, "well, he was outside working most of the time." His father was a cattle rancher and spent hours upon hours outdoors tending the cattle or doing maintenance on the property. His answer made sense on one level but it did not answer the bigger question – who was this man sitting at my table.

I asked again, "ok, but IF your father was in the house would he be sitting at the kitchen table drinking coffee?" To this, Jim replied, "well, if he was inside, yes, he would be sitting at the table drinking coffee." Then just like Carol

I Do Believe In Ghosts. I Do, I Do...

Anne right out of the movie Poltergeist I exclaimed, "He's here!"

After this encounter, Jim reported being visited by his dad frequently. Since that time he often told me how much he enjoyed time talking to his dad and how he would look to him for his counsel and advice.

Tamecia's Uncle

Tamecia came to see me one day. She wanted a psychic reading. Tamecia was a woman in her mid-thirties. She was curious about the direction her life was moving. We started the session with an aura reading. The aura is the electromagnetic field around the body. Much can be discerned about an individual by evaluating their aura. During an aura reading, a client can find out what is working for them. They also learn about where they are stuck, or not moving forward, as well as areas that may be limiting their growth.

At the end of the reading, I asked Tameica if she had any questions. She was interested in learning more about her spirit guides. I quickly turned my attention from her aura and began to focus on the energy of her spiritual allies.

Standing behind her, I saw the figure of a man. He was well built, muscular, yet short in stature. He had a kind and gentle nature. His energy seemed warm and loving. I got the impression he was around her a great deal of time.

There was something about him however, that did not feel quite right. His vibration did not match what I have typically experienced when interacting with spirit guides. I felt he had recently died and crossed over to the other side. I also believed that he had come back to this world to help and support Tamecia.

I described the man I saw standing behind her. He was a guardian angel of sorts. I let her know he was looking over her. I also made known that he might have been someone she knew in life. I provided her with details of his appearance. I used my body to depict the image of someone who was big and muscle bound. My client could not think of anyone who fit that description. I then conveyed to her that in comparison to herself, he felt older, like her father or perhaps an uncle.

Tamecia's eyes lit up when I made this comment. "When you said big did you mean tall?" she questioned. "No, not tall but big as in stocky" was my response. The smile that came across her face was priceless. She did have an uncle who had passed away many years earlier. She was in her teens at the time. He was her favorite uncle. She recalled how he, having no children of his own often doted on her.

Let me point out that I do not normally talk to the dead. This case was an exception. My communication with Tamecia's uncle was direct and unmistakable. He made it evident that he wanted me to pass along a message to her. He wanted her to know that he had come to her aid on a couple of occasions in the recent past. He made it clear that she would easily be able to recall the situations in which he intervened. To my surprise, she knew the exact circumstances he was talking about.

On one occasion she was driving in her car on the freeway when the traffic suddenly stopped. She believed she was doomed to crash. She instinctively swerved to the left in a desperate attempt to avoid a collision. Miraculously she came to a safe stop on the shoulder of the road unharmed.

On another occasion, she had stopped at a fast food restaurant with her two small children to get something to eat. As they left the restaurant, they encountered an unsavory young man. He began following them as they made their way back to her car. Tamecia was nervous, to say the least, and prayed nothing would happen to her and her children.

Suddenly the man lost his footing and began to fall. It took the would-be assailant a moment to regain his balance. That brief pause gave Tamecia the time she needed to get her children and herself safely into her car. Tamecia believed some kind of divine intervention had saved her and her children in both instances. It had! It was her uncle coming to her aid.

Uncle Bill

During an investigation into disturbances reported in a couple's home, I encountered a man who I called "Uncle Bill. He was found watching over Janice and Scott, the homeowners and their three-year-old daughter.

I sat down with Janice and Scott and described the man I saw at the conclusion of the investigation. His energy was kind and benevolent. He was a large man, not tall or muscular but overweight. His age led me to believe he could have been the father or uncle of one of the couple, as opposed to brother, cousin, or grandparent. Janice and Scott shrugged their shoulders and shook their heads as I described their visitor. They could not think of anyone who matched this description.

I probed deeper into Uncle Bill's energy. I believed he was associated with Scott. I was quickly told that Scott's father was still alive and he was also an only child. I dug even farther into Uncle Bill's personality. The words "good ol' boy" kept coming to mind. I also saw images of large Navy ships around him.

I shared this insight with the couple. It turned out that Scott's dad had been a Navy man and had graduated from Annapolis. I suggested Uncle Bill might have been a close friend of his father's. They might have shared a deep unspoken agreement to keep each other safe. I felt as if this commitment carried on beyond the grave. Uncle Bill was now watching over and keeping Janice and Scott's daughter safe.

Vortexes & Ley Lines

Vortexes and ley lines are often associated with paranormal or ghostly activity. Some people believe vortexes act as portals, which break through the fabric of our space-time continuum. A vortex is a spot on the earth, like Stonehenge in England, where great concentrations of energy are emitted from the planet. Ancient and indigenous cultures around the world revere these sites as sacred and have built monuments and temples to their gods on them.

These "power spots" are described as magnetic, electrical or electromagnetic. They are categorized based upon the energy experienced within them. A magnetic vortex, for example, attracts or draws energy to it while and electrical vortex energizes and activates the nearby area. An electromagnetic vortex carries the qualities of both an electrical and magnetic vortex combined.

Ley lines on the other hand are believed to be magnetic fault lines, which crisscross the earth. It is thought that ley lines act as roads and even superhighways by which ghosts can travel from one location to another. It is conjectured that movement along a ley line provides the energy needed by ghosts to manifest into our 3-dimensional world.

It is also held that when two or more ley lines cross, a vortex is created. Some compare vortexes and ley lines to the hub of a wheel. In this configuration, the vortex acts as the hub of the wheel and the ley lines act as the spokes that emanate from it. Another way vortexes and ley lines are viewed is from the perspective of Chinese medicine. A vortex can be likened to an acupuncture point and ley lines to the meridians that conduct life force energy from point to point throughout the body.

Regardless of which came first, the vortex or the ley line, it is understood among paranormal investigators and metaphysical practitioners alike that these anomalies are either rifts between dimensional space or the focal point of spirit energy.

Margaret's Apartment

Let me share with you how a vortex affected one woman who was having problems in her home. Margaret contacted Metroplex because she was having strange and unexplained things happening in her apartment. Every time she opened the door to a large closet in her bedroom she felt the presence of someone watching her. She also complained of ongoing problems with her computer and other electrical appliances in her home.

Margaret was a woman in her middle to late 30's. She had a professional career and while she believed in the paranormal, it was not something that would ever happen to her. Margaret moved into the apartment two months earlier and the problems she experienced started almost immediately. She tried to rationalize what was going on.

I Do Believe In Ghosts. I Do, I Do…

After a short while she was unable to come up with excuses for the strange disturbances any longer.

We started our investigation inside her one bedroom apartment. I immediately found the residual energy of an older man. He was standing inside Margaret's bedroom closet. I was able to detect his strong presence as soon I walked into the room and I could understand why she felt like she was being watched. She was! Creepy as his presence in her bedroom was, I assured Margaret he was of no danger to her.

I did not sense any other anomalies as I explored the rest of her apartment. There were no more ghosts lurking in the shadows. Instead, I felt dizzy and was becoming quite queasy. It felt like the floor was rocking beneath me. It reminded me of standing on the deck of a ship that was out at sea. Margaret's apartment was located on the second floor. I could not imagine how living in a second floor apartment could cause the rolling sensation I was experiencing. I did a quick scan of the people (live or otherwise) in the apartment below her and of her neighbors on either side. I did not discern anything disturbing coming from their units.

I stepped outside onto the second floor walkway to get some air as the group carried on its investigation. Still feeling a bit nauseous, I looked out into the neighborhood. Diagonally across the street was another building. It was also part of her apartment complex. To the far left side of the building I noticed the center of a huge vortex. The diameter of the vortex was so big the majority of her complex was encompassed by it. This explained why she was having problems with her TV and computer.

The energy of the vortex was very strong and unstable. I believed she was not the only one in the complex having problems with their electrical appliances. All of them were caused by the vortex. The instability of the vortex created other problems for the complex's residents. The vortex caused them to go through a lot of mental and emotional ups and downs. This can be difficult to navigate, especially if you are sensitive to energy.

When I went back in and visited with Margaret, I asked if turnover in the complex was high. She validated my suspicion. I shared with her what I detected about the vortex. I told her that everyone in the complex was being affected by its energy and whether they knew it or not they would choose to leave. The vortex's energy was too volatile and created too many inner changes. Margaret revealed to me that she had already planned to move once her 6 month contract was completed.

Jen & Rob's Home

Jen and Rob had recently purchased a new home and suspected they might be plagued with ghostly activity. They called Metroplex to perform an investigation. I made my way through their home and did not detect any anomalies – well, at first. When I walked from their kitchen into their dining area, my opinion changed. In the doorway between the two rooms I sensed something strange. It could best be described as a wall of energy.

The wall reminded me of something right out of Star Trek. The sensation I experienced could only be described as what I imagine it would feel like if you could walk through a force field like those used in the ship's brig. The brig secured its prisoners in their cells with a wall of energy, a force field. The force field was invisible to the naked eye, but if the prisoner tried go through it, any forward movement was obstructed. If however you could move through it I would imagine you would be able to feel its presence as you moved from the cell into the adjacent room.

At first I thought the energy wall ran parallel to the wall that separated the kitchen from the dining area. After studying it for a moment, I noticed that it was situated at a slight angle to the left. The bottom of the wall started deep within the earth and ran lengthwise through their entire house and out into the neighborhood. I also noticed that the wall seemed stronger in some areas as opposed to others. The area with the greatest intensity was in the doorway that separated the kitchen and dining areas. Then it went in their entry hall.

I asked Jen if the activity she and Rob observed occurred primarily in the kitchen and entry hall area. She answered affirmatively. She said the ghosts always appeared to walk the same path – through the kitchen into the hall and out of the house. There was something about the visitors which confused Jen. The ghosts she and her husband experienced always seemed different.

Based upon what I saw combined with Jen's comments, I concluded that their home sat on a ley line. I explained to the couple that ley lines are channels of electromagnetic energy and are believed to act as ghost superhighways. I let them know that ley lines can run close to the earth's surface thus spurring on paranormal activity. This is what I believed was happening in their kitchen and entryway.

This explanation seemed to make sense to the couple and answered many of their questions. Now Jen could understand why she sensed different people in her home. There were different people...a lot of different people! I reassured Jen that her invisible visitors were not there to cause any problems. They were just passing through and did not take notice of her or her family.

Portals

Portals are other points in which entities can make an entrance into a location. A portal is a door that can be opened into another time or dimension. Portals travel across the earth, some through time and space. Others are thought to open into alien worlds. Entities can also travel via a portal from one dimension or version of reality into another and then return.

Portals are a natural part of our physical universe. They can appear just about anywhere. They can also be created through the power of intention. Portals created through the power of intention are most often created by accident. For example, children are famous for accidentally creating portals. One of the great things about portals is that they can also be sealed, thus closing the doorway to wherever they may lead. We will talk more about these kinds of portals when we discuss attached entities.

The movement of a ghost or other entities through a portal is a type of haunting. We will explore this when we talk about the Titanic.

The Titanic

Metroplex was called to the home of Ben and Irene who reported some unusual phenomena. It did not take long before I discovered something anomalous in their home. In one of the back bedrooms I detected a portal. This was not a typical portal. Most portals are designed to allow "energy" to travel in and out of it. This portal, unlike the vast majority of them, only had energy moving in an outbound direction. This seemed very odd to me.

Continuing my evaluation of the home, I moved into the dining area. At one point I stood next to the family's china cabinet and found another portal. This portal, unlike the one in the bedroom was moving in an inbound direction. My curiosity was peaked. I wanted to explore this phenomenon further so I stepped out of the portal's energy field and stood in the middle of the living room. From that vantage point, I could mentally observe both portals.

What I detected was astonishing. Every two to four days the energy behind the inbound portal grew causing the doorway to bulge into the room. Once enough energy had accumulated behind the door, the portal would open. Once opened, a group of people would exit.

One regular visitor was a woman who wore a black full-length dress and a large black hat. Her hat had an oversized black feather sticking out its left side. Her dress suggested that she had lived sometime around the turn of the century and was financially secure. Associated with her was a woman who appeared to be of a lesser position in life. Her dress and mannerisms implied that she worked for the woman in the black dress. This woman seemed to be frantic. Her concern for a missing child was overwhelming.

I Do Believe In Ghosts. I Do, I Do...

A few men also made their way through this breach in the fabric of time and space, into the dining room of this couple's house. As a whole, all of the people who emerged from the portal seemed to be trying to escape a dire situation. Something dreadful was going on around them. Once through the portal their energy would linger in the kitchen for a day or two. Then, as if by clockwork, they would travel down the hallway and exit the house via the outbound portal in the bedroom.

I asked Ben and Irene if the events they noticed were periodic. I inquired if they had noticed increased activity for a couple of days and whether it would seemingly stop. Then, after a few days, the activity would repeat. This indeed was their observation.

As I continued evaluating this strange phenomenon I kept having visions of groups of people running around panic stricken. Images of the Titanic after it crashed into the iceberg kept flowing through my mind. It was difficult for me to tell if what I sensed was true or if the homeowner's home contaminated my impressions. Scattered around every room of their home were pictures and books about the Titanic. I asked them about their fascination with the Titanic. They commented to me that they had not become interested in it until after they had moved into this house.

I will never know if Ben and Irene were picking up information from their uninvited guests or if their interest in the Titanic tainted my perspective.

Stagnant Energy

There is another kind of energy that has often been confused with ghostly phenomena. This is stagnant energy. Stagnant energy forms when emotional energy fills an environment. It is not the same as a residual haunting where a traumatic event is imprinted on a location. Instead, it is an accumulation of emotional energies such as grief, pain, sorrow or fear. It builds up over time and penetrates the fabric of the environment.

One way to understand stagnant energy is to think of it as dust. We have all experienced how dust accumulates on items in a room. If we dust regularly we can keep any buildup to a minimum. If we only dust once a year, we will discover a thick coating of dust covering everything. The same holds true with stagnant energy. If the emotional energy in a location is not periodically cleared, it will begin to build up. Eventually it can begin to influence those who enter the affected space.

Most people can sense stagnant energy intuitively. It is often sensed as a feeling of heaviness. Sometimes individuals unknowingly notice the emotions associated

I Do Believe In Ghosts. I Do, I Do...

with the stagnant energy. They may find themselves weeping when they enter a room filled with sadness or they may feel agitated when they encounter the energy of distress.

These feelings may be perceived in a specific room or throughout a home. Believe it or not, everyone has encountered stagnant energy. Think back to a situation where you walked into a room and it felt to you as if the air was not moving. Did you detect a heaviness or even an oppressive feeling there? This is the feeling of energy that has stagnated.

On a number of occasions I have gone into a home believed to be haunted. What I often discover is a residence filled with stagnant energy. Remember, there are no otherworldly activities happening in these situations. Characteristically the homeowner will complain of feeling stuck. Their life might not be moving forward as planned. They may also believe they are not growing personally or spiritually.

Many times their physical environment will have unopened boxes stacked up in a corner, ones that have not been touched for months. They might have closets filled to the brim. The closets might be so full the residents avoid opening the door. I have also gone into homes that appear neat and tidy on the surface but could really use a deep cleaning.

If you suspect your home is plagued with stagnant energy, doing a house healing is a great idea. It cannot hurt and can only help. If you do not know how to do a house healing, keep reading.

Ghost Hunting 101

Are you interested in investigating things that go bump in the night? Ghost hunting is a fun hobby sweeping the nation. You can enter into the world of ghost hunting with as little as a hand held camera. If delving into the realm of the paranormal excites you, here is some basic information about ghost hunting that can help you get started.

Ghosts are electromagnetic (E/M) in origin. They create E/M fields when they manifest into this dimension. The energy that is given off by a ghost causes disruptions in the magnetic field near them. This energy emission allows their presence to be captured by specialized pieces of equipment.

Basic Equipment

Digital or 35mm Camera – Having a camera is probably the easiest and least expensive way to begin. Pictures, whether digital or recorded on film, can capture paranormal activity. This activity is often invisible to the naked eye. Ghosts make their presence known through the manifestation of orbs (believed by some to be the energy or the soul of an individual). Orbs appear as circles or balls of light in photos. Some orbs may be explained as dust particles reflecting light. Be aware of this effect and check for dust when taking photographs. Ectoplasm is another indicator of activity. Ectoplasm is a smoke-like or fog like mist that may appear in pictures.

Notebook & Pen or Pencil – As simple a tool as this may seem, it is always a good idea to have something to record your findings during an investigation.

Flashlight – Many times an investigation will take you to dark foreboding locations. These places may be a cemetery, old warehouse or abandoned building. Many of these sites do not have electricity. Having a portable light source will keep you from tripping over low profile tombstones, tree roots, trash or debris.

Extra Batteries – Depending on the paranormal activity of a location, batteries often mysteriously, and unexpectedly, lose their charge. This can leave you in the dark. If you are using any kind of equipment that requires batteries, you will be glad to have new ones to replenish the ones that may die.

Intermediate Equipment

EMF Detector – The Electromagnetic Field Detector is probably one of the most important pieces of equipment to own if you are serious about doing an investigation. In addition to detecting the energy fields of ghosts, it is a key tool to use when taking baseline readings of a site. A baseline reading is a preliminary evaluation of E/M levels in a location. It is recommended that investigators identify any existing or manmade electromagnetic fields such as those generated through power lines or electrical appliances. This should be done at the beginning of an investigation. This will help investigators identify E/M readings that are manmade in origin from ones that come from a paranormal anomaly.

Digital Video Camera – The video camera can be a useful tool. Unlike a still camera, a video camera will document activity in its entirety. It will document the length of time the phenomena occurred; the surrounding conditions as well as the phenomena itself. A tripod to rest the video camera upon is also helpful especially if it is going to be left in a stationary position.

Tape Recorder with an External Microphone – There is no better way to capture EVP's (Electronic Voice Phenomena) than through the use of a tape recorder. You can use a traditional tape recorder with high quality tapes or a digital recorder. An external microphone should always be used when attempting to capture EVP. A detached microphone will eliminate the sounds of internal gears that may contaminate your recording.

Advanced Equipment

Motion Detectors – Motion detectors can be used to sense the movement of unseen forces. Many brands of motion detectors need to be connected to an outlet. Battery operated detectors are also available from some manufactures. Decide which type of detector will best support your investigatory needs before you decide to buy one.

Thermal Scanner – A thermal scanner, or non-contact thermometer, can be used to detect rapid temperature changes. A change of ten degrees or more from the ambient temperature (the temperature of the surrounding area) can indicate an ethereal presence.

K2 (or K-II) Meter – A K2 meter is similar to a

I Do Believe In Ghosts. I Do, I Do...

standard EMF meter. EMF meters use a needle to indicate electromagnetic fields. K2 meters use a series of LED lights to indicate changes in the electromagnetic field. It is alleged that when a ghost is present it will disrupt the electromagnetic field around it. Many ghost hunters are using this device in an attempt to communicate with the ghost. They hope that the ghost will interrupt the localized electromagnetic field to indicate the answer to yes or no questions.

Using A Psychic As An Investigative Tool

In addition to the cameras, meters and gauges, used by ghost hunters, many groups also utilize the services of a trained psychic. Psychics, when used correctly, can provide depth and breadth to an investigation. Unlike the pictures, videos and audio recordings gathered at a location, information collected by a psychic does not need to be reviewed. They can provide the property owner or the other investigators instant feedback. This is the role I take when working with Metroplex and other ghost hunting groups.

Finding a true psychic can be a real asset to the quality of your investigation. The psychic can play an invaluable role in supporting the efforts of any paranormal investigation. Like a fine tuned piece of equipment, the psychic can help detect anomalies as well as provide additional insights into the nature of activities encountered.

If your group decides to include the use of a psychic as part of the investigative team, there are a few items that should be considered. To maximize the objectivity of the investigation, it is critical that the psychic go into a location "cold." This means that the psychic does not know the details of the location or about any activity that may have been encountered by other group members or the property owner. Telling the psychic the history of a property can cause the individual to become "contaminated." Contaminating the psychic can make it difficult for the psychic to differentiate what he or she has perceived from what the psychic has been told.

Investigation Basics

Once you have obtained a basic equipment kit you can begin to explore the world of paranormal phenomena. There are a few ground rules to bear in mind before you start an investigation. First, never go on an investigation alone. In addition to personal safety, you never know what may happen. Next, consider the fact that ghost-hunting groups are often judged by their behavior and professionalism. Always be respectful of the people, places and situations you may encounter.

The first step in any investigation is to select a location. Locations can include cemeteries, an old building, historic sites, private homes and businesses. Always get permission to be on a site. This will keep you from getting in trouble for trespassing. Some paranormal groups do research on a site's history before beginning a formal investigation. Talk to the property owner, check out old newspapers or contact the local history society. Ask if anything "interesting" has happened there in the past. It is also a good idea to evaluate the location for potential hazards. You can also use this time to identify spots where you might want to place stationary video cameras, motion detectors or tape recorders.

Arguably, most paranormal infestations occur after dark between the hours of 9pm and 6am. These hours are considered by some to be the "psychic" or "witching" hours. It is believed to be the best time to record paranormal activity. If a ghost does take up residence at a location I personally found that it is there all the time and not just at certain times of the day.

The time has come to start your investigation. Begin by walking around the location and getting a feel for the surroundings. Set up stationary equipment and take baseline EMF and temperature readings. Then let the investigation begin. Take pictures, shoot video, record for EVP and take meter readings. Take them everywhere and anywhere. Especially if you feel something or if you get a reading on another piece of equipment. Turn your stationary equipment on and let it run when you have completed walking through a site.

When you have finished collecting your data it is time for an analysis. Here each picture, video and tape content is reviewed for evidence of a haunting. This is probably the most important part of any investigation. Be skeptical as you review your materials. Look for earthly causes of any recorded abnormalities. Was someone smoking in the area? Was the room or location dusty? Were there reflective surfaces that could cause the abnormality in the image? It is important to make sure your evidence will stand up to scrutiny by eliminating any other explanations. This will give your data more credibility.

There is a downside to ghost hunting. You will quickly find not every location has ghostly activity. You may discover that many of the things you initially detected during your investigation have a real world explanation as opposed to a supernatural one. What keeps many ghost hunters going is the one-in-five chance that this investigation will be the "real" thing.

When your analysis is done, present your finding to the home or property owner. Many groups post the results of their investigations on their group's website. This allows them to share the outcome of their investigation with the rest of the world.

If you are interested in ghost hunting, there is a surplus of information available on the worldwide web. I'm sure you will be able to easily find articles about ghost hunting in general, or stories of other ghostly encounters. You can even find places that sell ghost-hunting equipment. With the increase in popularity of ghost hunting you may be able to find the websites of other ghost hunting groups in your area.

Sons Of Herman Hall

Metroplex was contacted by a local television station and asked to participate in a news segment they were putting together about ghosts. The piece was to air on Halloween. We were to meet at the Sons of Herman Hall in Dallas, Texas, to perform an investigation. Built in 1911, the "Hall" has been a gathering place for the "Sons of Herman," the nation's oldest fraternal benefit society. It was also used to host musical events, meetings, weddings and other celebrations.

Every room of this hallowed location was filled with residual energy of its prior guests. There were "people" in the hallways. There was a "young man" in the small bowling alley, and a "woman" in the back office. On the second floor there was a "group of young middle school girls" who could be seen leaning over the strong wooden railing. They were whispering and giggling amongst themselves as they kept a close eye out for any cute boy who might make his way up the stairs.

The entire second floor consisted of one large ballroom. I was surprised by the lack of activity, residual or otherwise that I experienced as I walked around this space. It was when I stepped up onto the small stage in the far left corner of the room that my attitude changed.

Once on the stage, I turned to look out into the open room. A man, who was surrounded by a semi-circular group of other men, was standing in the middle of the room. I honestly could not tell you what they looked like. Each of the men I saw was encased in a shroud of brilliant white light. They were feeding the man in the center energetically. At least that is the way it seemed to me. He was the focus of their attention. They were sending their energy to him. I was amazed by the intensity and sheer power of the energy they emitted.

I watched this event with fascination as it unfolded before me. For some reason Vicki and Wayne, both members of Metroplex, walked right into the area I was observing. I had not said a word about the men I saw in the middle of the room To everyone's surprise their E/M meters spiked.

Vicki and Wayne slowed down their investigation. They took some baseline readings in other areas of the room to double check the accuracy of their equipment. Then they moved back into the area where their E/M meters went wild. This time, they were also working with a non-contact thermometer to check for temperature fluxuations. They were able to trace, using their meters, the exact outline of the half-circle I was observing. Every time they measured an area where one of the men stood, they registered a drop in temperature and a spike on the E/M meter.

It is not very often that what I see clairvoyantly is

validated by ghost-hunting equipment. In this instance it was verified to my great satisfaction.

Finding Ghosts Intuitively

Everyone has the ability to detect ghosts intuitively. Perhaps you have sensed something around you but did not realize it was a ghost. There are a number of ways in which ghosts can be perceived. You just might "know" they are there, what they want and why their energy can be found at a particular location. On the other hand, you might actually see them, hear them or feel their presence.

For most people "feeling" a ghost is the easiest way to perceive the vibration of someone who has passed. A great indicator when a ghost is nearby is when the hair stands up on the back of your neck and arms. Tightness in the chest, or in the pit of the stomach, can also act as a sign. Especially if the ghost has a negative or heavy feel to it. Experiencing your skin tingle, vibrate or sensing a temperature change (from hot to cold or cold to hot) can also indicate something is present.

I was giving a lecture on psychic abilities at a ghost hunting conference in Jefferson, Texas. On this trip a good friend Wayne joined me. Although supporting my work, he is in his own words "a skeptic at best." I decided to have him meet Bruno while we were there. You may recall meeting Bruno when we talked about the Jefferson Hotel earlier. I could tell from Wayne's demeanor that he did not expect to experience anything. To him, all of this ghost stuff was just a bunch of woo-woo, airy-fairy B.S.

After we entered the hotel, I had him walk partway up the stairs to the second floor. That is right. I had him stop about ten steps from the top of the staircase… just outside Bruno's energy field. Then I had Wayne climb one more stair. I could feel his energy begin to shift as he encountered Bruno. So I asked, "Are you noticing anything?" To his astonishment he replied, "my hands are tingling and the hair on my arms are standing on end." I never told him that hair standing on end is an indicator of a ghostly presence and was thrilled when he reported his current condition.

After a few seconds, I had him take another step up, a step deeper into Bruno's energy field. With a look of, I do not know if it was shock or just plain disbelief; Wayne reported that the sensations were stronger. Slowly he climbed yet another step. The strange feelings he described earlier now fully surrounded his body and were seemingly even more intense than before. What he did not realize was that he was standing right smack dab in the middle of Bruno's energy. I had him stand there for a few moments so he could fully absorb the feeling. By the time he reached the top of the stairs all of the tingling, vibrations, hair standing on end feelings Wayne was experiencing earlier, had stopped as mysteriously as they had begun.

My face beamed with pride as Wayne made his way down the stairs. I looked at him with loving eyes and said, "Welcome to my world." In return, he smiled brightly, a look of excitement on his face and said, "Let's do it again!"

The Dr. Rita Method

In the course of my work many people ask, "How do you do that?" Well do not tell anyone but I will share some secrets of how I detect ghosts with you. But please, let's keep it between you and me.

For me, successfully finding ghosts intuitively is about paying attention to (and evaluating) what I am seeing, feeling or sensing. The first thing I do when arriving at a potentially haunted site is to stand outside the home and take a few deep breaths to clear my mind. This helps to open me up to anything I might encounter. Entering the home I move meticulously through each room with my hands held out slightly in front of me. I use my hands like antenna to detect any changes in the environment. A change in energy can be very subtle, so paying attention to even slight variations is important.

My hands will often tingle when they enter into a spot where an anomaly is present. I may detect resistance in my hands or a shift in pressure. Sometimes I will feel sick, queasy or nauseous. When I first started this work, I thought the queasiness may have been caused by something I ate. I have learned over time that it is an indicator of heavy duty (usually negative) residual energy or ghostly activity.

When I sense something around me, either in my hands or fully in my body, I stop in order to get a better idea of what I am feeling. Much can be learned about an entity solely by its feel. Using my hands, I start assessing the anomaly. My first task is to identify the boundaries of what I am noticing. How big is it? How far does it extend into the room? Are its borders hard or are they soft or wispy? Does the energy feel good or bad, positive or negative, emotionally warm or cold and detached?

We can also learn a lot about a ghost and delve deeper into its emotions by checking in on (evaluating) our bodies and emotions. Ask yourself "How do I feel now?" What are my feelings and emotions telling me? Am I feeling happy, sad, protective, concerned, remorseful or angry? If you walked into the room happy and filled with excitement and you are now feeling depressed and suicidal you are most definitely sensing the emotional energy contained in the location.

At times I will physically step into the center of a ghost's energy field and allow the emotional energy to fill me. This is what I did during the investigation at Sheppard's home. I have stepped into the ghost's energy body and opened myself to feel what they are feeling. This may sound strange but it really does work.

The Baker Hotel

An example of my intuition on overdrive was when I was at the Baker Hotel in Mineral Wells, Texas. I was part of a large investigation that was being filmed for one of the cable networks such as the *Discovery Channel*. In its heyday, the Baker Hotel was a hot spot for those of affluence and influence. Many famous figures, from politicians to movie stars, were reported to have passed through its doors and enjoyed its luxurious accommodations.

The hotel had been abandoned for many years. It was without electricity or running water. The halls and many of the hotel rooms were littered with an odd assortment of old broken pieces of furniture mixed with garbage. The only thing I knew about the hotel prior to arriving was that it had been a mecca for the rich and famous. I was also told about a "health spa" located on the second floor of the

I Do Believe In Ghosts. I Do, I Do...

facility. Even before we entered the hotel, I was inwardly drawn to investigate the spa area as opposed to the many other rooms that comprised the building.

Now, in my mind, when I think of a spa, I envision hot tubs and saunas, whirlpool baths, fun, excitement and maybe a little mischief. This is not what I found at the Baker.

Overlooking the lobby of the hotel was an open balcony. On the balcony was a very angry male ghost who was pacing back and forth across the room. I was told we would have to walk through the balcony area to get to the spa. My first thought about the prospect of coming in close proximity to the man was an overwhelming sarcastic, "great!"

As soon as we moved through a set of double doors that separated the balcony from the balance of the second floor, I immediately became nauseous. Not in the nauseous, heavy way I described before but as if I was actually coming down with something. This queasy feeling persisted as I walked in and out of the many rooms that fanned off the main hallway. Depending on where I was, or what energies my hand came into contact, the uncomfortable feeling would increase in strength. Once I moved on... it would start to fade away.

With a flash of insight, my inner knowing kicked in. It was now clear to me what the Baker Hotel was really all about. At the turn of the century, "health spas" were very common around the United States. They were places where people would go to restore their health and vitality. Health spas during this period were built on (or near) natural springs that were purported to have some sort of healing properties. Thus, the name of the city "Mineral Wells" made total sense to me. The Baker Hotel was not just for the rich and famous. It was here that people from all walks of life came with hope of saving themselves from some devastating disease or illness. It was the residual feeling of death and dying, or those weak in constitution that I felt as I walked through the long forgotten halls of the spa.

Types Of Abilities

I utilized two different psychic abilities at the Baker Hotel. The first was discerning what was going on around me by paying close attention to the sensations of my body. I needed more than feelings if I was to understand what was really going on in the spa. To do this, I opened myself up to receiving information on other sensory levels. In this case it was my knowingness or clairconsciousness.

Knowingness is an intuitive skill where you just "know" the answer. Knowingness is like having a flash of inspiration. Where it comes from, and how you know what you know, can often seem a bit mysterious. Information received through our knowingness tends to be overwhelmingly accurate no matter how far out it may seem. Information received on knowingness levels always come through unedited and unfiltered. This is what happened at the Baker.

It was later in the investigation when I learned that the water in Mineral Wells has high levels of lithium (a drug often prescribed to individuals with bi-polar disorder). Individuals who came to the spa would drink and soak in its healthful waters. More often than not, they would miraculously begin to feel better. With a reputation like that... I could understand why the weak, sick and infirm would sojourn to the hotel. In some instances the water was not enough to save their lives.

Another way to detect a ghost is by seeing them

intuitively. When talking about ghosts, most people write off their experiences as being a figment of their imagination. They believe the only way you can see a ghost is through a full-blown apparition. To this day, I have never seen an apparition in any of the homes I have lived in or locations I have investigated. I am not sure what I would do if I happened to encounter one, but running the other way screaming does sound about right.

People typically see ghosts in two ways. The most common way is out of the corner of their eyes. Then when you turn your head to look at them they have mysteriously disappeared. The ghosts in this kind of sighting are referred to as shadow figures. Shadow figures are ghosts that appear as a dark outline or the shadow of a person moving across a room. Shadow figures are elusive and difficult to detect. It is hard to tell whether they represent actual spirits or are residual in nature.

I discovered a shadow figure living in the backyard of one of the homes I owned. His name was Tom. At least that is what I called him. Tom was the previous owner of the house. I could only assume Tom enjoyed working in the yard. It was filled with small manicured bushes and shrubs that centered on a birdbath of two lovebirds in a sweet embrace. It was something right out of a Good Housekeeping magazine.

For the longest time I thought I was the only one who would "see" Tom as he slipped by the large, floor to ceiling windows that overlooked the back yard. That is until one day when one of my sons wanted to know who the man was in our backyard. To my surprise, everyone in the family had seen Tom at one time or another. Many times friends, with a concern that our house might have been broken into, would become alarmed when they detected movement in the yard. With a smile on our faces we would just say, "Oh no, that's Tom."

Ghosts can also appear in full living color and not just as a shadow that scampers across a room. This is the way I see ghosts. I see them in my minds eye, clairvoyantly. You might wonder how I can differentiate between what I am seeing in my mind's eye as opposed to my physical eyes. I can tell because I can still see the ghost even when my eyes are closed. "Looking" with my eyes closed provides me a greater visual range and an increased level of detail than when my eyes are open.

Everyone has the ability to use their inner vision to both visualize images in their mind's eye as well as to access information clairvoyantly. We are not taught to pay attention to these subtle messages. More often than not, we invalidate them, attribute them to coincidence or toss them out as a figment of our imagination. And while I'm mentioning this as we discuss clairvoyance, the same holds true for all of our intuitive gifts.

When we see a ghost in a room, we merge the input we are receiving through the physical eyes with what we are obtaining clairvoyantly. Using this kind of seeing when you walk into a room, you are able to see what's in it and any ghosts that may be in the vicinity.

Another question I am also frequently asked is if I regularly see dead people. They are curious to know if I see their discarnate souls walking the halls in the hospital or at the mall. For some psychics this may be the case. Thankfully, my gift does not work that way. If I want to see into the world of the unknown I open myself up to receiving that information. It is similar to putting a pair of psychic glasses on. With them on I can see auras, presences from the spirit world, and all manner of intuitive information. When my work is done I can easily take them off.

There have been countless investigations where upon first entering a location I would comment, "oh there is a man standing over there. He looks like..." Or "there's an older woman walking around the room over there." This does leave me to wonder what people are actually seeing when they "see" an apparition. Perhaps they too have accidentally put their psychic glasses on and are seeing with a combination of their physical eyes as well as with their inner vision.

Regardless of the method you may use to detect a ghost, it is always fun and interesting to delve deeper into their story if possible. This is done by asking questions of the ghost and opening yourself up to receiving an answer. Each of us can access information on knowing levels (clairconsciously), visually (clairvoyantly), auditorially (clairaudiantly), or on feeling levels (clairsentiently).

It is through our clairaudience that we may be able to hear what a ghost has to say. Clairaudience refers to our ability to hear spirit. We are all familiar with our inner voice. It is the voice we hear inside our heads, which reminds us what to do. We argue with it, ponder things with it or work through problems with it. Like our inner vision, where we can see spirit as well as visualize, when we receive messages via our clairaudience we may confuse it with our inner voice.

Each of our intuitive abilities provide us with different types of information, assuming we are open to receiving it. A good friend of mine Carol Layman, from Journeys Between Lives, calls these abilities, channels. We all have a primary channel by which we receive intuitive information. Some people primarily use their visual channel while others use their hearing, feeling or knowing channels. One is not better than the other, but is how our body has learned to process information. In addition to our primary channel, we can also learn how to tap into the information provided by the other intuitive channels by giving them a little time and attention.

Having our impressions validated greatly helps speed up the process. It lets us know we are truly receiving intuitive information and reassures us that our impressions are correct. From my perspective, many times it feels as if I am making it all up. It is the shocked look, the head nod, the big smile or the affirmative comment that lets me know I am on target. Validation helps us to separate truth from fantasy and fact from fiction.

Mr. Baker

At the same investigation in the Baker hotel I described earlier (in fact on the same floor) I walked into one of the many rooms that filled the second floor. Standing looking out the window, was an older gentleman. A man perhaps in his sixty's, he was tall but stocky in build. He had receding white hair and wore a clean pressed white button down shirt and dark tan slacks. On one side of the room I saw a desk which he would pace to pause and then move back to the window. I received this visual information clairvoyantly.

Feeling into his energy clairsentiently, I could tell that he was someone who had a high position within the hotel's management organization. He bore an air of privilege and superiority, as if he looked down upon the people who worked for him, or for some reason, he kept himself separate from them. I felt responsibility weighing heavily on him and it seemed as if there were a number of problems he was trying to resolve. Problems that were gnawing at his stomach. He looked out of the window as if pondering what to do. It felt like he was carrying the weight of the world on his shoulders. I assumed that he was a doctor or the spa manager. I recall thinking "this guy really has a chip on his shoulder."

Something about the man kept bothering me after the investigation ended. I tried to ignore the gentle reminders spirit offered. Finally giving in to its prompting, I went online and did some research about the Baker Hotel. Looking for pictures taken at the hotel, I hoped to catch a glimpse of the man I saw in the room. To my surprise, the man standing in the was Mr. Baker the hotel owner!

Dark Angels

Baby, I'm Stuck On You

> *"If your cat's speaking Latin, you might have a problem."*
>
> – Jason Hawes

People are usually unaware of the presence of ghosts, angels or spirit guides around them. It is also reasonable to assume they are not aware of non-corporeal beings that may be tied to them. Most people view attached entities as the stuff of a good horror movie, or novel, and most certainly not something that can be influencing us and our lives. Startling as this may be, 20% to 30% of all people suffer from an entity attachment.

Attached entities, the majority of the ones we will be discussing in this section, take on a physical form. They present themselves as having human bodies. The ability to take on human form indicates these entities have lived here on the earth. It also implies they "speak English." I do not mean in the literal sense. However, since these entities have incarnated on this planet we share similar points of reference and the same cognitive reality. This makes communication with these entities possible. This is not the case when talking about alien entities or demons who fall into a category all their own.

What Are Attached Entities

Unlike angels and spirit guides who support us, or ghosts who may come to visit from time to time, attached entities take their interaction with us one step farther. They penetrate the auric field.

The aura is the electromagnetic field around our bodies. It is our personal space. Our aura can also be described as a boundary which defines limits. It can be likened to a border that marks where we end and another begins. It is what separates us from others and it works to help identify who we are as individuals. It also acts as our first line of defense – physically, emotionally, mentally and spiritually. It helps us to be clear about who we are, what we think and how we feel.

Think about a time when an uninvited individual stood too close to you. I will bet it made you feel uncomfortable. This person was standing inside your auric field. The feeling of physical discomfort you experienced was an indication that your energetic boundary had been violated. We are often consciously aware of this kind of violation and even more so when we are touched or hugged without permission. The feelings of anxiety or uneasiness we experience develop because our boundary is being crossed. Our personal space is being invaded.

An individual's emotional energy can also enter the auric field. When we take on another's emotional energy, we typically react to this intrusion subconsciously. We experience this unwanted energy as a part of ourselves instead of recognizing it as an intrusion. We think, believe and interact with the thoughts, feelings, needs and desires we sense are our own. We assume we are the ones who must deal with them. We take responsibility for what we perceive within ourselves and endeavor to satisfy the need or resolve the problem.

New car shopping, for example, often leaves us vulnerable to someone else's energy. Has this ever happened to you? You go to a car dealership just to take a look at the new cars. Just for fun, you take a car out for a test drive. The highly motivated and energetic salesperson directs your attention to all of the outstanding features of the wonderful car you are driving. The salesperson points out how smooth it rides, how well it handles, and brags about the unbeatable gas mileage. He may have you recall the inadequacies of your current car. Then he may ask you to notice how good it feels to drive this brand new car. Then before you are done, there is the coup de grace. "If you act now…" adds a sense of urgency to your already saturated energy space.

Everything feels so good to you as you drive the new car. You get excited about the wonderful opportunity being presented. You feel compelled to act. You think, "I have to have this." Then before you know it, you have taken out your checkbook, signed contracts and are the proud owner of something you may or may not have wanted in the first place. Once you leave the dealership, the energy of the sales person dissipates from your auric field and you wonder what the heck you have just done to yourself.

I will bet there have also been times when you have tried to handle a problem and found that the issue only rolled around and around in your mind to no avail. No matter how hard you tried to solve the problem you were unable to come up with a solution. I would suggest that in these situations, the problems you were trying to resolve really were not yours.

We collect energy from other people, including their troubles, all of the time. We are especially susceptible to this among our friends and family. Once their emotional energy has penetrated our aura, we respond to it by trying to fix the problem. Unfortunately, as is often the case, we cannot fix it, heal it or resolve it because it was not ours in the first place.

Like the emotional energy of others, we can also allow an entity to enter our auric field. When an entity penetrates our aura, it is classified as being "attached." Angels, spirit guides and ghosts do not enter the auric field. They respect our boundaries and honor us and our personal space. Attached entities, on the other hand, do not care about boundaries and believe it is perfectly okay to come into our energy field. This lack of respect for us is the reason why "attached entities" are put into a lower, more harmful, category than ghosts. Even if a ghostly presence is mean, vindictive and spiteful, he or she is still honoring us and our personal space.

I firmly believed, in the past, that these non-corporeal beings were uninvited guests who had somehow invaded our space. Upon closer examination of individuals suffering from entity attachments, the overwhelming majority invited them in unintentionally. We will be discussing this concept in detail as we move further along. For now it is important to recognize the significance of having one of these beings in our auric field.

The first question most people ask when talking about attached entities is, what do we experience when an entity has entered our personal space? It can be summarized into two words – control and manipulation.

All attached entities are parasites. Their vibration is low. They are typically ego driven, emotionally negative and out for themselves. When they are in our auric field, we experience their thoughts, feelings and emotions, however nasty they may be. The vast majority of people with an attached entity are unaware of its presence. The sufferer assumes they are responsible for creating the ongoing patterns of negative thoughts and undesired emotional responses they are experiencing. They end up owning the entity's energy as their own. They eventually believe "this is who I am." They mistakenly live their lives surrounded by these lies and feel powerless to change.

Are you wondering if you have an entity attached? There are a number of signs that can indicate if one is present. Symptoms of an entity attachment include:

- A history of physical, emotional or sexual abuse
- Disassociation or being ungrounded
- Memory problems
- Hearing voices or an inner voice that constantly criticizes you
- Repeating patterns of behaviors
- Anxiety or panic attacks
- Irrational bouts of fear, anger, sadness or guilt

- Sudden changes in behavior or mood swings
- Depression or thoughts of suicide that you cannot seem to stop
- Addictive behaviors, including addiction to alcohol, drugs, cigarettes, sex or gambling
- Impulsive behavior or an attraction to dangerous situations
- Illnesses that will not respond to treatment or are of an unknown cause

Granted, everyone can have a bad day. It has been my observation that if any of these issues are ongoing or persistent in your life, then it may be a good idea to find out if an entity is plaguing you.

Loren

Loren came to see me for a medical intuition evaluation. As the session began to unfold I asked her to describe her chief complaint. Bulimia was her reply. Bulimia is a disorder characterized by excessive eating followed by compensatory behaviors such as fasting, vomiting and/or the use of laxatives (aka, purging) and over-exercising. In her mid-thirty's, Loren, upon first glance, did not appear to have weight issues. Loren could lose a few pounds, or maybe exercise some more, but she was not fat or in any way obese.

I had never worked with someone with bulimia. My curiosity was aroused. I asked Loren what she ate daily. To my utter astonishment, her response was "Six Big Mac's." I countered her statement with "Is that with the fries and the cokes?" I can still feel my jaw dropping when she let me know it most certainly was. I then inquired if the six Big Mac's were spread out throughout the course of a day.

No... This was an example of what she would eat for lunch.

Still fascinated by her tale I asked what she ate the rest of the day. A dozen eggs with toast or a couple of boxes of cereal for breakfast was her reply. Apparently after gorging herself on the excessive amount of food, she would promptly go into the rest room and quickly regurgitate her meal. We never talked about her evening dining fare but I was convinced she had a problem.

Bulimia is not presently viewed as a physiological disorder. It is not a dysfunction of an organ, gland or tissue. Since her problem was not physical in nature, it required a different approach to assess her problem. My goal was to help Loren restore her health. I was not quite sure how I was going to accomplish this but I was going to give it my best shot.

As I contemplated where to start, Loren made a very interesting comment, which changed the course of the entire session. She told me that years earlier, a psychic told her she was possessed. At first, I did not think anything of it. It is almost routine to hear reports from clients of some pretty far out stuff psychics tell them. For some reason, as I began to evaluate her, I was unable to dismiss the notion of her being possessed from my mind. Instead of fighting it I decided to go with it. I put my "entity glasses" on and changed my focus from evaluating her subtle body to look for entities. I was amazed at what I saw. There was indeed an entity attached to her who apparently was very, very hungry. He was hungry all of the time.

We spent the remainder of the session working on moving the entity out of her auric field and helping it take its next step. We both said goodbye as it exited my office and stepped into the light.

Approximately two weeks later, I received a call from Loren. She reported her eating habits had miraculously returned to normal right after the session. What she thought was her problem alone really was not her problem at all. It was the impact and influence of the entity in her space that had been the real culprit.

Entity Motives: The Truth About Entities

Dealing with an attached entity can be like interacting with the power and influence of a religious, social or political cult. These cults may look wonderful at first, but the reality is all they want from you is your obedience. Cults will often use trickery, manipulation and mind control to win you over. They work by telling you how happy or successful you will be if you follow their guidance. They will hide the truth of who they really are and what they actually want. They will only make their hidden agendas known when they think you are ready. Then once their hold over you is complete, they will use coercion, guilt, character assassination and even intimidation to keep you from leaving their influence.

Entities do the same thing. They will lull you in with their promise of a better life, protection, strength and security. Once inside and firmly attached the game changes. What was once promised goes by the wayside. Where they once provided you with good positive thoughts and ideas you are now filled with negative ones. They will tell you over and over again how awful you are, how unloved or how worthless you are. These negative tapes can keep playing in a never-ending loop over and over again in our heads. This is done to control you and keep you in a negative or altered emotional space. While this interaction may be detrimental to you, for them it is desirable. And the deeper, darker and lower you go and the longer they can keep you in this emotional place, the happier they are. Relief can only be experienced if we can manage to move the entity out of our space. This is when we can reestablish our sense of self.

Many times an entity will not reveal its true motives when confronted. They will silence themselves and will not want to divulge its hidden agenda. In my mind, their silence says loud and clear they are out for no good. It is not that they cannot tell you or do not know. To the contrary. They will blatantly say they are not going to tell even if their reply is through their deafening silence.

It has also been my observation that all attached entities lie. They always try to make it appear as if they are looking out for you even if what they are telling you is how bad you are, how good another drink would make you feel or by providing you with guidance that ultimately is not in your best interest.

In some instances, the communication an individual receives from their entity will tempt the person into doing things such as cheating, lying, stealing or even worse. The entity may also tell the person how to get away with the misguided deed. It will often help the sufferer get out of trouble if caught in an act of indiscretion and will provide them with more lies, deceptions and excuses they can use. The entity will do whatever it takes to get both of them out of a jam. This can mislead the individual. They believe they are receiving good and helpful advice from their entity.

From the sufferer's point of view, they never get caught, never get in trouble and never have to take responsibility for their actions. To them, life is good, so why change.

A lie, specifically if it will keep them out of trouble, is much easier than telling the truth and dealing with the consequences of the deeds. What they do not realize is they are being deceived into believing they have no choice but to be this way. They think lying, cheating and being out of their integrity is their only option. This is their truth, their reality, even though it is severely distorted. If they only knew this is not their true self.

Cords: How Entities Attach & Control

In addition to being in our auric field, many entities will create an energetic cord between themselves and one of our chakras. The word chakra is a Sanskrit word that translates as wheel or disk. Chakras are said to be energy centers or vortexes we use to send, receive and assimilate energy from the world around us. Like the auric field, they are part of our subtle body. There are seven major chakras located along the spine. They start at the base of the spine and travel up the body to the top of the head.

Each chakra processes a different type of information. They include:

Chakra/Ability	Definition
First Chakra	
Survival	Our ability to take care of our body; i.e., food, shelter. Houses our fears and insecurities.
Physical	How the physical body interfaces in the world.
Health	Our current level of health.

Chakra/Ability	Definition
Second Chakra	
Clairsentience	Our ability to feel energy.
Desires	Our ability to recognize our needs (including our sexual needs).
Creativity	Our ability to create for ourselves.
Third Chakra	
Willpower	Our personal power, will power, drive, motivation.
Passion	Our ability to have passion in our lives.
Out of Body Experience	Our ability to dream; astral travel.
Fourth Chakra	
Self-love	Our ability to love ourselves unconditionally.
Affinity	Our ability to love others unconditionally.
Fifth Chakra	
Clairaudience	Our ability to hear others; i.e., our spirit guides.
\Inner Voice	Our ability to listen to our own spiritual information.
Telepathy	Our ability to communicate with others without speaking.
Sixth Chakra	
Inner Vision	Our ability to see mental image pictures.
Clairvoyance	Our ability to see and interpret psychic information.
Seventh Chakra	
Knowingness	Our ability to just "know."
Trance Mediumship/Body	Our ability to leave the body.
Trance Mediumship/Being	Our ability to allow another being into our body.

Cords are energy channels that connect one point to another. They can be compared to a wormhole that cuts through the fabric of time and space. A more rudimentary way of thinking about cords is to picture them as the string that travels between two tin cans allowing you to talk to your friend.

Cords allow energy and information to be transmitted between two people. They keep the lines of non-verbal communication open between friends, lovers and even our enemies. I have never met anyone who did not have at least a few cords attached to them.

A classic example of a cord in action is in the instance of a new mom. The infant, dependant on his mother for survival, will often create a cord between its first chakra, its survival center, and its mother's. When the child is in need, such as when he or she is hungry or has a soiled diaper, it will send a message to its mother via this cord. This connection can be so strong it can awaken its mother out of a deep sleep.

Communication cords can also be created between an entity and ourselves. In most instances, the entity will attach to one of the upper three chakras of the body, that is either the fifth, sixth or seventh chakra. The fifth chakra is responsible for our communication space. Through it we receive information non-verbally on audio channels. The sixth chakra holds our inner vision and our visual space. It is also responsible for how we see the world and where we form our reality and belief systems. The seventh chakra supports our ability to receive inspiration and divine guidance. It is also identified as the location where we, as spirit, can enter and exit the physical body.

When a cord is formed between an entity and one or more of these centers, the entity has a direct connection into our being. The clarity by which we experience the entity's lies and distorted truths reinforce our belief in their reality. We see the world through their unrealistic view of it and we assume it is true.

Oma

One day I had a session with a woman from India named Oma. Dressed in a beautiful golden sari she was interested in an aura reading. When performing an aura reading each of the seven layers of the auric field is assessed. The client is told what is working for them or where conflict is being detected in their psyche.

Sharing with her what I saw in the first layer of her aura, I quickly discovered it did not really matter what I had to say. From her point of view, everything I communicated was wrong. I tried to not be affected by the constant and repetitious invalidation I was experiencing in the session. I continued on, layer by layer. It was when I finally got to the sixth layer of her aura that I discovered the problem.

Firmly attached to her sixth chakra was an energy cord. On the other end of this well-established cord was an entity. Without exaggerating, the cord had to have been the size of a baseball in diameter. The cord went directly into her sixth chakra and affected how Oma saw the world.

She was unable to see life, in particular her life, from a clear and neutral place. Her perceptions were filtered, distorted and warped by the entity's tight control over her. It was no wonder why she was unable to identify with anything I told her. What I shared with her I believed in my heart to be true. It was the entity who did not want her to perceive any reality other than the one he provided her. It was the entity who invalidated me and the session as a whole.

I never said anything to Oma about her attached entity. My impression was that she was in total agreement with having it there. I also felt this entity had been in her personal space controlling her for a long time. To remove it without her full agreement would have been unethical. Its beliefs and dictates were so fully integrated into her life that to remove it would have changed the world as she knew it forever.

Jason

My son CJ, a number of years ago, had a close friendship with a young man named Jason. I believed Jason was having a negative influence on CJ's behavior. Jason had also befriended one of my girlfriend's son. Tara did not like the influence and control this young man exhorted on her son either. It was our observation that whenever our sons spent time with Jason they would return home acting mean, aggressive, self-righteous and arrogant. Now granted, they were teenagers, but we both could see a noticeable change in them whenever this young man was around.

My friend and I are both psychics. We decided to take a look at Jason's energy. We wanted to evaluate his motives and rationales. We were interested in finding out how and why he influenced our children so greatly. It was not long into our reading of Jason's energy field that we both detected the presence of attached entities. These entities would not only prompt his misbehavior they would also cord the boys – or perhaps more correctly stated – enlist Jason into cording the boys. This would on an unconscious level, control them, their thoughts and their emotional state.

One day we decided to do a joint intervention with our sons. Trying to remain calm so that the encounter would not turn into a shouting match we began our discussion from an educational perspective. Both boys were brought up around metaphysical and new age topics. We knew that they would have some understanding of what we were talking about. Nevertheless, for the sake of clarity we started at the beginning.

"Do you know what entities are?" we asked. To which they both replied they did. We had them explain what their definition of what an entity was, especially an attached entity. Their reply was perfect and fully to our satisfaction. We then shared with them our belief that Jason was affected by a number of entities attached to him. We also let them know that his entities were influencing their behavior for the worse.

Before we had a chance to continue, CJ jumped into the conversation. He let us know that he was aware of his friend's entities. In fact, according to CJ, Jason had three entities attached to him. To our surprise he was able to rattle off their names. He also let us know that Jason was well aware of his entities. Apparently, he worked with them and communicated with them on a conscious level. This scared Tara and me. It is bad enough being unconsciously controlled by an entity. To befriend it and allow it to guide you into mischief... that is really bad.

What they did not know was communication cords had been created between the entities and themselves. The cords were plugged into various parts of their bodies and chakras. They also did not consider that maybe, just maybe, the negative really funky attitude they were walking around with was not actually theirs. We also told them that we believed that unless they became free and clear of Jason's influence they were going to end up getting into trouble. Trouble was not a place we wanted to go. They agreed.

We then taught the boys how to remove the cords

from their bodies and subtle energy space. I think they were both relieved when they were able to feel themselves in their bodies without the influence of their friend. I know that when I looked at them after the healing session, they both looked a bit spent but peaceful.

Why Entities Attach

Entities attach to us for a number of different reasons. Most importantly, entities attach to us because we open the door and invite them into our life. In most cases, it is accidental and unintentional on our part. Of all of the people affected by attached entities, the majority create this symbiotic relationship as children. This is especially true of children who are sexually molested or experience long-term physical or emotional abuse.

It is common for the child to feel alone and abandoned by those who were supposed to take care of them, protect them and defend them. In their hurt, pain and solitude, they call out for someone to help. They project a wave of emotional energy out into the universe like an SOS or a beacon of light. They ask, hope and pray for someone to come and save them. The depth and strength of emotional energy they project is so intense their wishes do get answered. By whom? Well the entity, of course.

Yes. It is the child who opens the door and allows the entity to enter. As a way of coping with the dysfunctional situation, the child disassociates from his or her body. This leaves room for the entity to step in. Unfortunately when talking about an entity... a little bit can be a lot.

As the entity exhorts its control, in a weird way, the child receives a reprieve from the painful circumstances they are experiencing. They might take on an "I don't care" or "You can't hurt me" attitude. This creates a boundary around the tender and gentler parts being violated. They may become angry, spiteful or filled with rage. They may totally disassociate from their bodies leaving the entity fully in charge. This can cause the child to move forward in life without any memories of the offending behaviors they experienced.

It is hard to determine in these situations if the entity is good or bad. Entities that attach to children are drawn to them because they are broadcasting their emotional wants and needs. The angry, defiant, rebellious or rageful energy the entity infuses into the child helps the child cope, although dysfunctionally, to the turbulence around them. In turn, the entity feeds on the negative, pain filled vibrations of the child and the circumstances they are struggling to endure. So while it is not a great situation, the entity does help the child survive mentally and emotionally – well, relatively speaking. As they say, something is better than nothing and I guess this is something.

In the beginning, the entity may have helped the child cope with what was happening to them. As the child matures the reactions offered by the entity to a potential threat often no longer serves them. They may begin to see their overstated response to a situation as being inappropriate and may want to embrace a new way of interacting with the world. Thus comes the rub.

In most cases, attached entities do not occupy the auric field all of the time. I say in most cases. There are some individuals who are constantly being bombarded with an entity's influence. We will discuss this later on when we explore mental health issues and attached entities. In the majority of people with entity problems the entity will keep a short distance away. This provides the individual a measure of autonomy over their bodies, thought processes and emotions. That is until the entity is called in again.

Children and adult children call out to their attached entities through their emotional space. It is believed each emotion carries with it its own unique vibration or frequency. Love, joy, serenity, contentment, anger, frustration, sadness and grief all oscillate at different rates. As subtle beings, we are all able to sense the emotional energy being projected by an individual. We can all learn to convert felt senses into words through a little practice and experience. An intuitive, for example, can detect and identify the emotional energy of a person, place or situation by tuning into, and interpreting, the specific vibration they feel.

Entities are also able to sense emotional energy. If a specific emotion is experienced by the affected child, such as ones that trigger a past hurt, the entity will come in to save the day. Oh yeah, and it will try to take control of the individual again. There is a direct connection between the individual and their entity. It can be likened to the bat phone which has one big button on it. A direct dial to the entity. Once the call is made and the entity is back in their auric field the individual will experience a response that, while it may be out of proportion, will seem familiar, safe and consistent.

This is what I find fascinating. As we grow, evolve and work to heal our past, the buttons we used to call the entity begin to diminish. It might be months or even years before something happens that sets off a response within us which calls the entity. Once back in our aura it will begin exhorting its influence over us again. You guessed it. It will help the affected individual find that deep dark ugly place within themselves. Once there, it is difficult for the individual to figure out what they want or need.

George

I will bet you are wondering, "How does she know so much about attached entities? The answer is quite simple – I had one of my own!

Unbeknownst to me, "George," a discarnate low vibrating soul had been negatively influencing my emotional wellbeing for years. George was an attached entity who breached the security of my auric field and filled my mind with negative, self-deprecating thoughts. I can only assume George came into my life when I was very young. Growing up was challenging at best. I honestly do not remember a time when things got bad that George did not show up and exhort his control and influence over me.

He was not always in my energy field but when he made an appearance I would be filled with anger and frustration. His influence caused me to be defiant to an extreme. I knew that regardless of what happened to me or around me I would not be hurt. I was led to believe it was a sign of weakness if I let others know I had been upset or affected in any way. I remember countless times as a small child staring directly at my Mother, into her eyes, with a look of contempt on my face.

It was not until I was an adult that I learned about George and his affect; not only on my behavior, but on my whole life. In retrospect, I realized that when George entered my aura it was as if a dark cloud had engulfed me.

It manifested as premenstrual syndrome (PMS) on overload. I would be critical, cranky, withdrawn and pessimistic. If someone said or did something that bothered me, the chance that I would go off on a tirade was high. This fog of negative existence could last for hours, days, or even weeks. Then something would happen. I might hear a silly joke. And just as fast as "my mood" came on, the clouds would separate and the sun would shine in my life again.

It was many years after learning about George that I was finally able to move him out of my aura – permanently!

Addictions

Addictions, especially drug and alcohol abuse are another common reason why entities attach. The altered states these substances create within the body can be a calling card inviting entities into the auric field. When we are intoxicated, we naturally put down our boundaries and drop our guard. This makes it easy for an entity to enter into our auric field. They can have greater control over us when our senses are impaired.

These entities often want to live vicariously through us. They are looking for one more drink, one more high or another encounter. We often hear stories of people having drastic personality changes when they are under the influence. When an entity is involved, a nice clean cut, calm and humble man may be instantly transformed into a raging idiot. Just add alcohol. Some individuals, after a binge, report they just were not themselves. After they sober up they end up regretting their inappropriate actions. Others have no recollection of what they did or where they were while under the influence.

Once the drugs or alcohol wears off, it is much more difficult for the entity to enter their aura and exhort its control over them. The entity will then try to talk the affected individual into participating in the addicting behavior again. The entity works hard at keeping the object of their compulsion on their victim's mind. The individual may think (or perhaps obsess) about their next opportunity to indulge. They are reminded just how good it felt, how their troubles will go away or how it really does not matter anyway. "So go ahead...just have a little one for old time's sake. It will be okay."

If you think about it, if these individuals were clean and sober how could the entity attach to them? It is in the entity's best interest to keep them addicted.

Richard

A mutual friend referred Richard to me. Richard believed his life was falling apart. His job was in jeopardy as well as his 11-year marriage. What was once working smoothly was on the verge of ruin. Why? Because Richard had a drinking problem.

In my opinion, I did not believe seeing his life fall apart is what motivated Richard to call. It was the fact that a few days earlier, while in a drunken stupor, he looked into the mirror and saw two men staring back at him. When he turned around to look at the men standing in his bathroom, no one was there.

Things really began turning for the worst when he told his wife about the invisible intruders. Scared for her children and her own safety she called 911. Richard was immediately taken to the hospital and put under a forty-eight hour psychiatric watch. What made matters worse was the fact that Richard had already gone through the drug and alcohol rehabilitation program offered by his employer. The terms of continued employment stated he needed to remain drug and alcohol free. How could he explain this unexpected and unexplainable absence to his boss? Richard could not tell him that he had been so drunk he had started seeing things and had just spent the last two days locked up in the "loony bin." Afraid of what was now moving well out of his control, Richard called James, our mutual friend, who suggested he call me.

During our first session, Richard revealed to me his long history of drinking. He also disclosed that he had been hearing voices for years. He had never shared this small detail with another living soul. Without even looking at Richard's energy I knew we were going to be dealing with entities.

With his tale fully told, I looked at Richard's auric field. There were indeed entities attached to him. There was also a large vortex going into the back of his aura. I shared my observations with him. We spent the remainder of the session discussing the implications of having an entity in his space. He scheduled another session so that we could do some intense work on moving his entities out.

We began clearing out one of his two entities during our next session. This entity's name was John. He had attached to Richard when he was a child. This entity had a really nasty disposition. It was obvious to me that John had influenced Richard throughout his life. It was John who would perpetually talk Richard into some kind of mischief. If caught, John would help him out of it by providing him with lies to tell or a finger to point. This worked well for much of Richard's life. He rarely got into trouble. The downside of this relationship was that Richard never learned to take responsibility for his actions.

John also influenced Richard to begin drinking. Over the years, Richard was taught how to cover his tracks. He learned how he could remain under the influence of alcohol without anyone detecting it. Well, until recent days.

It took a while to get John out of Richard's auric field. I was not sure at the time if John had fully moved out of Richard's life or if what I was being shown was an act to get me off his back. It was not long before John was back and took up residence in Richard's aura once more. As a result, Richard was drinking again, unemployed and in the process of divorcing. I guess the consequences of his actions had finally caught up with him.

Accidental Attachment

Long before man began documenting his achievements in written form, the secrets of man and his relationship to the universe were being taught. It was believed that the concepts and information being transmitted from teacher to student was so powerful that if left in untrained hands the result would be disastrous. The high priests of ancient Egypt are known to have initiated select individuals into their sacred mysteries. In Judaism, the study of Kabbalah (the basis of Jewish mysticism) was only granted to men, over the age of forty, who had a good understanding of the Torah and Talmud. "Mystery schools" were created during Greco-Roman times to transmit knowledge about the inner nature of God to the indoctrinated.

It is only in recent years that "occult" (esoteric, metaphysical or new age) principles have been made available to the public. Formerly, those who explored its secrets where highly trained in their concept and use. Warnings and precautions have always surrounded its use and for a good reason. When dabbling with seemingly benign tools, techniques and methodologies... a door can be opened, a welcome mat can be placed out and a sign can be posted inviting an entity into our personal space.

Ouija boards, for example, fall into this category. Calling spirits to you is the whole purpose of the Ouija board. Popularized in the mid-19th century by the Moderns Spiritualist Movement, Ouija boards are said to allow the user to receive messages from the other side. Ouija boards transitioned from a tool of the spiritualist when Parker Brothers mass produced its version of a Ouija board. This allowed anyone to venture into the realm of the supernatural.

A Ouija board, also known as a spirit or talking board, is composed of a flat board with letters, numbers and a set of symbols covering its surface. It is accompanied by something called a planchette. This device is used to point to individual numbers and letters on the board. The numbers or letters that are selected are joined to answer questions or form messages.

Strong warnings exist regarding the use of a Ouija board, yet individuals still play with them hoping to receive divine guidance from the spirit world. What many people do not realize when using a Ouija Board is they are potentially opening the door to the other side. If done without precaution, you never know what might arrive. This includes negative entities who have an agenda of their own.

The use of techniques such as automatic writing or even just calling in our angels and spirit guides can also leave us vulnerable to an entity attachment. It has become popular for people to ask their guides and angels for support and guidance. They are typically not specific about who they are summoning. When their "guide" does appear, they do not evaluate its energy. They are just happy someone has shown up in response to their request.

As you may recall, attached entities are attracted to lower vibrating emotional energy. If you call upon "your guides" when you are scared, depressed, frightened or sad you can accidentally attract a negative entity who can ultimately cause more harm than good.

I have heard stories of entities attaching to individuals from many different sources. Their tales are essentially all the same. When the entity first started working with the caller, the messages, communications, insights and advice they receive feels divinely inspired. The good feelings the entity provides will trick the individual into trusting it (remember entities can and do lie). Mistakenly the caller may let their boundaries down. This can invite the entity in even further. As time goes on… the entity messages start to change. What was once helpful, uplifting and useful now becomes messages that can provoke fear, intimidation and control. Some callers, since they believe this communication is coming from a higher source, may act upon the messages received without discriminating the pros and cons of what they are receiving.

Once we submit to an entity's control what happens next is unpredictable.

Meredith

I worked with a woman named Meredith who decided to try her hand at automatic writing. Automatic writing is a form of channeling in which messages are said to come directly from the spirit world (we will talk more about channeling later in this section). This form of divination calls for the practitioner to allow messages from the other side to flow through them. Information received in this trance like state is documented in written form.

Meredith had been a student of metaphysics much of her adult life and had been working with automatic writing for a number of years. She believed automatic writing would be an effective way to gain insights into herself and her life. As she became more proficient with this technique she began to hear a voice around her. Her early successes excited her. She naturally assumed it was one of her spirit guides talking directly to her. She encouraged the communication with her guide thinking, "it was all good."

As time went on however, the voice changed. It went from the sound of one guide to the voices of many. The style of communication she first experienced also began to change. The words she heard were originally kind and uplifting, now they were cruel and demeaning. Music and an array of odd sounds soon followed adding to the mix of voices she was already hearing. "I can deal with the tone of their communication," she confessed, "what is bothering me the most is they talk all the time!"

The first thing I saw when I looked at Meredith's energy was a group of nine entities in her auric field. They seemed like the souls of individuals who may have come to her for help. They did not seem to be causing any trouble although they were adding to the sense of heaviness Meredith felt. These entities did not seem firmly attached and I knew I could help Meredith move them out of her auric field quickly and easily.

In addition to these nine entities there was a smaller group of four entities. These four entities were loud and obnoxious and appeared to be the ones responsible for many of the problems Meredith was experiencing. Unlike the other group who I believed I could easily help

Meredith move out of her personal space, this group did not seem interested in going anywhere. They seemed to be having too much fun in Meredith's aura.

But that was not all. Standing behind Meredith was a big and powerful entity who seemed to be the ring leader of the second group. He appeared to be the most negative and vocal of them all. He was firmly implanted in her energy field.

Hiding in the back of Meredith's aura I also detected two large vortexes and five smaller ones. The open door of the vortexes allowed entities to come and go from her energy field undeterred. No wonder why she kept hearing strange new voices all the time.

We began work to heal Meredith's aura during our next session. I was easily able to help Meredith move the first nine entities out of her auric field. Once they were gone we began working on the ringleader. I believed that once he was out of her space the others would quickly follow. We started this process by trying to communicate with him. We wanted to find out what he wanted and why he had chosen to attach himself to Meredith. In a nutshell, we were told that he was not going to talk to us. He also informed us that anything he did say would be a lie.

With this information in hand, I went to plan B. One of the things I noticed when interacting with him was he had inserted a large cord into the back of Meredith's neck. It went right into the back of her 5th chakra, her communication space. Working together, we removed the energy cord. Now his hold on Meredith was not as strong and we were able to move him out of her auric field. We spent some time helping him work through the "issues" that kept him bound to the earth and assisted him as he made his way into the light.

We then began the process of sealing up the vortexes in her auric field. This proved to be a bit challenging but we did make some progress and sealed up a few of the smaller portals.

Meredith seemed much relieved at the beginning of our second session. She stated that the volume and tone of communications she was receiving had settled down to an almost bearable level. Meredith and I are still working on helping the remaining entities transition. I believe after a few more sessions we will have her energy space free and clear of all her entity attachments.

A Word Of Advice

I would like to offer a bit of advice to anyone wanting to work with tools to communicate with spirit. First off, do not take interacting with the spirit world lightly. When you do establish communication, be specific. Be discerning. Use judgment before opening the door to an entity and trusting it fully. Since entities vibrate at different levels you will only want to bring in a being whose energy is high or full of light. Ask to speak with a particular angel such as Michael or Raphael or an ascended master like Jesus or Buddha.

Once communication is established, reflect on how their energy feels deep inside you. Each person, living or dead, carries a specific vibration and consciousness about them. When interacting with an entity do you feel good and uplifted or are you concerned, anxious or skeptical? Trust what your body is telling you.

It is also important to reiterate this point. If you are in a bad emotional place, you never know what you might attract. You might bring in an entity that is vibrating at a higher level than you at that moment, but if you are down and in the dumps how high is that?

Past Life Attachment

Entities can follow us from one lifetime to another. Outlandish as this may sound, it can and does happen. A "contract" can be formed between two individuals that does not end at death. It transcends the lifetime they spent together and moves forward through time and space. It is not uncommon, for example, for an entity to attach to someone out of a sense of loyalty. The person may have saved the entity's life (when he was still alive). The entity could have been a soldier in the past and still carries a sense of allegiance to his commander. These examples represent the good side of an entity attachment – not that any kind of entity attachment is good or desirable.

Entities that attach to us from our past lives may have a sense of ownership over the affected individual. They may have owned us or were so obsessed with their relationship with us, that they could not bear the thought of not being around us. Their desire to keep us, to own us, to not share us with anyone else can be so strong, their grasp so tight that it extends past the grave. This need for control over us is so great that it follows us from one lifetime to another where it works to possess us once more.

Zahi

Annette came to me suffering from a number of physical problems including migraine headaches, facial swelling and sever fatigue. In addition to her physical troubles, Annette had also been plagued by intense panic attacks that were affecting her ability to function.

In one particular session, we decided to explore the underlying cause of her panic attacks. As I tried to discover what emotional issues were triggering these intense reactions in her, the image of her father appeared in my mind's eye. It was not the first time during the course of our work together that we discussed how his overbearing need for control created an intense fear in her. This fear was so deeply felt it caused Annette's body to react violently.

Every time the image of her father emerged in session we would talk about the dynamics between the two of them. We would work to identify which of her many emotional buttons were being triggered and what were the associated false beliefs. My goal was to help Annette reclaim her strength, power and autonomy and support her as she released the unresolved emotional issues that were affecting her physically.

For some reason I happened to make an offhanded comment about Annette's father's bald head. This comment sent the two of us in a whole new direction. What I discovered was her farther, in this life, has a full thick head of hair. Who had I been talking about in past sessions? If it was not her father, who was it? Now we were both wondering what was going on.

I reevaluated the situation. I knew her father was involved with this bald headed man. I also felt as if he, the bald headed man, was in one way or another running the show. His energy was not only controlling my client but upon closer inspection, he was also controlling her father. Yes, it was a true family affair.

I asked the entity to step out of Annette's auric field. We asked him his name. He was not forthcoming. Every time I looked at him I was reminded of Dr. Zahi Hawass, the Undersecretary of the State for the Giza Monuments. For lack of a better name to use, since he was not being cooperative, we decided to call him Zahi.

The details of Annette and Zahi's relationship were convoluted at best. Annette last interacted with Zahi a few hundred years earlier when both Annette and Zahi had physical bodies. Zahi was a man of power, wealth and influence, who lived somewhere in the Middle East. His home was a tent which was filled with all of the luxuries a tent could have in those days.

During this lifetime he would often call upon Annette to have relations with her. I could not tell if he owned her, as in she was a slave, or if she was one of his wives. What I do know is that he lusted over her and had his way with her whenever he desired. Annette did not reciprocate the feelings. All she experienced in her exchanges with him was terror. If she could have run away, she would have. There was one thing she knew without question. If she did run away he would find her and kill her.

I shifted my focus from Zahi to Annette's dad. From this new perspective, it was easy to see that there had also been a relationship between Annete's dad and Zahi. Annette's father was a quiet man in this former lifetime. He had little say in any aspect of his life. Yes, Zahi controlled every facet of it. The pressure exhorted by Zahi was so great, the threat was so strong, he found it easier to comply with Zahi's wishes than fight. He put his own feelings aside and chose not to resist him in any way.

It was not clear if Annette knew or interacted with her father in this lifetime but it was apparent to me that Zahi knew them both. The control Zahi had over the two of them in the past extended far beyond their former lives. United in the present as father and daughter, Zahi's influence over both of them could still be felt. In this lifetime, her father was kind of a mousy man, not that you would know it. What people experienced when they interacted with Annette's father was Zahi.

Based upon the level of unquestioned authority and complete control Zahi had over Annette's dad, Zahi must have attached to him early in his life. Her dad was not connected to his emotional space. Her dad frankly did not know himself at all. I guess it really did not matter. In this lifetime, as in the former one, Zahi controlled it.

This is the intriguing but sad part of Annette's story. Growing up, regardless of the actual physical face she saw when she looked at her dad, all she could "see" and experience was the energy of Zahi. This made complete sense to Annette. She recalled even as young child she was afraid of her dad. Tied to this was a remnant from her lifetime with Zahi – a belief that if she did something wrong or went against him, he would surely kill her.

Family Entities

Another category of entity attachment is that of family entities. A family entity is a being who attaches itself to an individual. When the individual becomes too old, too physically weak or dies, the entity will move on and attach itself to another family member. By the time an entity is ready to occupy its next victim's energy space, it will often skip a generation. This is because the elder is typically still alive and in relatively good health when their sons or daughters are growing up and coming into their maturity. When the time does come for it to move on, the grandchild is seen as being young and full of vitality – Ahhhh... the perfect person to occupy. Family entities can be passed down through both sexes. However, I have seen it occur most often in woman. The entity will travel down the female line for generations.

Brenda

Brenda came to me for a session because she was experiencing a lot of confusion in her life. Strong, capable and competent, it seemed to Brenda as if her life was turning upside-down. Things she had once strived for now held no interest. She felt stuck and did not know what to do.

What I think concerned her most were the changes she was experiencing in her mood, emotions and attitude. In her words, she could not understand why she was being so "bitchy." She confessed that she did have moments of crankiness in the past but this phase, in her mind, had been going on entirely too long. She was looking for help and wanted assistance in moving this dark cloud out from over her.

It was easy for me to see that Brenda was not going crazy once the session started. She was not all that "bitchy" in her normal waking life. In my professional opinion, what was plaguing Brenda was not her attitude but an entity. Yes, she had an entity problem.

The first thing I noticed when we started the session was a woman standing in her auric field. The energy the woman carried reflected all of the characteristics Brenda had recently discovered within herself. Probing farther into the woman's energy, she seemed very old. Not in the age sense of the word. My opinion was based upon the way she was dressed. Her clothing indicated she had last lived in the distant past.

I asked the entity when it first attached itself to Brenda. I was a bit surprised to find out that it had only been within the last few years. Brenda did not appear to have a drug or alcohol problem and she did not seem like the type that would be playing around with Ouija boards. I asked the woman where it came from. The answer I received astonished me.

The woman seemed to be very familiar with Brenda. Brenda's energy reminded her (the entity) of the last person she was attached to. When I traced back through the entities history on a timeline, I could see that she had most recently been attached to another member of Brenda's family, her grandmother, on her mother's side.

I started to ask Brenda about her grandmother. I wondered if she had recently passed away. To my surprise, she was still alive. She explained that a few years earlier her grandmother had been very ill and was hospitalized. Although she was alive, she had never fully recovered from her illness. Her illness made her too weak to satisfy the needs of the entity. The entity moved on. It moved on right into my client's energy field.

At first, Brenda did not believe what I was telling her. She honestly could not fathom it was true. I asked if the way she was feeling reminded her of her grandmother. Almost embarrassed she admitted it did.

I shared with her that although not common, family entities do exist. I felt that this entity had been with her family since the late 1600's, where it moved through her family for generations. I was not able to determine why it had attached to her great-great-great-great-grandmother in the first place, but Brenda was relieved to know that "she" was not the problem. She was also grateful a simple solution to her troubles could be achieved.

Incubus and Succubus

You would never think that sex could come into play when talking about entities, but it does. Called either an incubus or a succubus it is said that all they want from their victims is sex. A succubus is an entity who takes the form of a beautiful woman with the intention to seduce and have sexual intercourse with men. This interaction typically occurs while the unsuspecting man is dreaming. According to some myths, the succubus's intention is to steal a man's seed. Others say succubi draw energy and vitality from men in order to sustain itself.

Its companion, the incubus, is an entity who takes male form in order to lie upon its victims, namely women. Their aim is to have sexual relations. According to some legends, a succubus will engage in sex with a man in order to collect his semen. The semen is either given to an incubus or the succubus transforms itself into its male counterpart an incubus. The incubus would then have intercourse with a woman. This was done in order to impregnate her.

I am not sure what to think about the potential offspring of an entity/human union, but I do know that incubi and succubi are real.

One Man's Pleasure

I have a quick and cute story about one man's reaction to a succubus in his aura. I was doing a reading for a middle-aged man at a psychic fair in Bellevue, Washington. What the reading was about, at this time, I could not tell you. But throughout our 15-minute session I kept seeing a band of energy moving up, down and around him.

I tried to ignore it but my curiosity finally got the better of me. I tuned into what I kept seeing moving around this man's body. It was a luscious, full figured, woman. She pressed her body firmly against him. Then she would caress him or reposition herself into another seemingly sexual embrace. Had she been visible to the naked eye, I'm sure someone would have suggested they get a room.

Astonished by what I saw, at the end of our session I commented on the woman who was obviously in his auric field. With a sly, sideways smile on his face he replied, "I know… isn't it great?" I just shook my head, rolled my eyes and smiled.

Fred

I first encountered Fred as part of a medical intuition evaluation I was performing. Amy came to me because she was experiencing full blown panic attacks that were interfering in every aspect of her life. She wanted to understand why she was having them and could not have been clearer in her desire to have them to go away.

I took a quick look at her energy field. The first thing I detected was an accumulation of emotional energy sitting right in the middle of her chest. I have often seen emotional energy, such as pain, loss, sadness, grief and even withdrawal trapped in this area of the body. Fear and anxiety can also become trapped here. When it does, it can manifest as panic attacks. In Amy's case, the emotion I encountered went well beyond fear. It was complete and utter terror.

Amy could relate to my observation. Terror seemed like a good way to describe what she often experienced. I began to follow this line of thought and I wanted to know when she had encountered something that scared her so badly. I looked at a timeline of her life to find the earliest point in her life where she could have experienced an event that evoked such a large emotional reaction. The age I received was between the ages of eight and ten.

When I asked her about this time in her life, she could not recall anything out of the ordinary. She did reveal that she had experienced feelings of terror at an even earlier age. When I looked at the timeline again, I did not detect anything abnormal in her younger years.

It occurred to me that perhaps the emotions she was experiencing were coming from a past life. To my rational mind, this would explain why I was not able to find an origination point. I suggested she see a hypnotherapist who specialized in healing past life traumas. I quickly concluded this part of our session. I did not want to contaminate her conscious mind with speculation on my part. This could interfere with her ability to discover the underlying cause of her problem.

Our session continued...

Amy asked if she had been sexually molested when she was a child. She suspected the abuse might have come from her stepfather. I had her say her step-dad's name so I could find him energetically. I did not detect anything out of the ordinary about him when I evaluated his energy. Her step-dad, although somewhat emotionally distant, seemed like a proud, caring man who would never think of doing anything like that to her.

She went on to describe some vague memories of being touched and having someone lying on her in bed. She assumed it was her step-dad doing these things to her – things she had blocked from her conscious mind. Then Amy made a few additional remarks about her experiences. It was her comment about feeling him stand by her bed at night watching her that caused my ears to perk up. I finally gave in to my suspicion and took a quick peek to see if it was an entity causing her problems. In my mind I thought "oh no, not another entity person."

I was not fully prepared for what I saw. Not that what I saw was gross or disgusting. It had more to do with what he was doing and how his energy felt. I summed up what I saw into one word and the first one that happened to come to mind that day was pervert. Yes, Fred was an incubus.

The image Fred shared with me was amusing in a weird and distorted kind of way. Instead of standing by her or behind her, like I have seen in the past, he was bent over at the waist. His intention was to look up her dress. He was wearing a wrinkled trench coat, crumpled clothes and appearing quite unkempt. He could best be described as looking like a homeless person. Fred also fit the stereotype of someone who would flash you as you walked past him on the street.

I asked Amy if she had ever heard of something called an entity. Fortunately, she had and was intrigued to know more. Then I dropped the bombshell on her. "Well," I told her, "you have an entity attached to you and the entity you have is a freakin' perv."

We spent time talking about entities in general and how we can be affected by them. In this instance I had to include information about her incubi. I suggested that perhaps she had confused the energy of the incubi with that of being sexually molested by her step-dad.

She was in a bit of disbelief. To help prove my point, I asked, "Do you dislike wearing dresses?" She would always feel uncomfortable if she wore one and until a relatively short time ago would never wear them. I explained to her my impression regarding her dislike for wearing dresses. "Fred likes looking up your skirt and had been doing so for a long, long time. While you might not be aware of his presence, your subconscious mind certainly was."

Because of the ongoing trauma issue that I felt had come from a past life, I referred her to Carol Layman from Journey's Between Lives. Carol specializes in helping clients resolve past life issues that are affecting them in the present. I suggested that she also find out if Carol could help her deal with who we were now calling Fred while they were in session. I was not sure if Carol would be able to help her resolve this problem but it was worth a shot. I told her if it was not resolved, she should visit with me again.

The next week I heard from Amy. Her session with Carol was not as successful as we had both hoped. Her intense feelings of fear and anxiety interfered with her ability to relax and find that place inside herself where the magic could occur.

Coming to my office the following week we set off on the task to get rid of Fred. I then began the process of helping him move on. We will come back to Fred in the next section when we discuss Clearing Entities.

Entities And Inanimate Objects

Entities can also take up residence in inanimate objects. I have a girlfriend named Gail who sold crystals at a psychic fair I produced. She would often talk about how this crystal or that one had a little spirit in it. I always thought her comments about beings inhabiting her crystals were just a bit out there. But the truth is even though I thought Gail's comments were strange I have to admit I too have some items with entities in them.

The ones that comes to mind most readily are "The Girls." The Girls are a set of three Franklin Mint porcelains. They are Power, Destiny and Fortune. I laugh about it now but after purchasing these three pieces (they were all part of one collection) I bought another porcelain figure also from the Franklin Mint. This one was of an Egyptian female goddess. I thought it made a wonderful addition to the three I already had.

Well...they (The Girls) did not like her. They did not want to be displayed next to her. They did not even want her in the same room. At one point, I wanted to purchase a curio cabinet in which they, along with some of my other treasures, could be displayed. Oh My God! Picky, picky, picky and their taste? Expensive! There were more cabinets they did not like than ones they thought would be appropriate. To this day, I still have not purchased a curio cabinet. Maybe I should blame it on them.

Anyway, let's get back to entities. The entities in Gail's crystals and the ones in The Girls all vibrate at a high frequency. They were not filled with dark, low level entities. But this is not always the case.

The Haunted Doll

On an investigation with Metroplex, we visited the house of a woman whose daughter had been drawing what she thought were ancient symbols and other figures on the walls of her restroom. A practicing Pagan, Megan did not believe she had ghosts but she knew something was not right. She was on a mission and was actively ruling out anything, which might explain her daughter's strange behavior.

Issues like this will typically not bring Metroplex out on an investigation but Megan sounded so desperate we agreed to go. Because of the bizarre nature of this investigation, I was informed in advance of the problem. It was hoped that I might be able to shed some light on the issue at hand.

Doing a cursory walk through Megan's apartment it felt as if she had done an excellent job protecting her home, thus keeping the neighbor's and any ghost's energy out. I told Megan I did not detect anything abnormal in her home. She directed me to the bathroom where her daughter had drawn the strange symbols. The symbols looked Rune-ish in nature. Were they some ancient text? I do not know. It was when she directed my attention to the bathroom sink that my interest in this case was ultimately peaked.

On the counter in front of the sink her daughter had drawn a pair of eyes. I swear those eyes followed me around the room. They made my skin crawl. Peering back at them, I got the impression of a man. Actually what I saw was the image of a pirate. His energy felt conniving and sinister.

By this time, her daughter Allison was standing in the doorway. I could tell from the look on her face that she knew exactly who this man was. I began asking Allison about the man whose eyes appeared on the bathroom sink. She told me that he was a pirate. She also shared a few other details about him. After a short period of time I could tell that what was once truthful information was now becoming the imagination of a six year old. I believed she had called him in and was communicating directly with him. This was definitely not good.

I sat down with Megan to discuss what I had discovered in her apartment, including the pirate in the bathroom sink. I also informed her that her daughter was indeed communicating with him. I suggested she find someone who could work with her daughter and her extraordinary gifts. With her daughter being so open to receiving energy, without proper guidance, she could go over to the dark side.

Megan then asked me about the doll which sat on a shelf in Allison's bedroom. I had not noticed it earlier and went back into the child's room to see it for myself. Chills ran up and down my spine as I looked at the doll. In my mind, all I could hear was a singsong voice saying, "I've got a secret, I've got a secret." That was enough for me and

I quickly exited the room and returned to where I was sitting with Megan.

Megan asked me what I thought about the doll. I repeated back to her in the same singsong voice I heard coming from the doll, "I've got a secret, I've got a secret." Megan's hand went to her mouth. She was shocked. There were many days when Allison would come to her, a mischievous look in her eyes and sing those same words. This scared Megan. She wanted to know what she should do with the doll. I suggested strongly that she give it away, throw it away, but most certainly get it out of her house.

I do hope Megan took my advice and got Allison some help. Most importantly, I hope she got rid of the haunted doll.

Alien Entities

So far, we have discussed many types of entities that take on corporeal form. The final classification of entities we will be exploring are what I refer to as alien entities. What separates an alien entity from the other types is that alien entities do not take on a traditional human form. Alien entities can assume a wide variety of shapes and sizes. I have seen alien entities that have a humanoid look to them. They have a head, a body, two arms and two legs. I have also seen alien entities appear as animals, insects and slugs. In one case they looked like giant amoeba or jellyfish like creatures that were attached to the skin of my client and were eating away at her flesh. That one was pretty disgusting.

Like your basic attached entity, alien entities come and go from the auric field. Their influence can be seen and felt but it leaves the individual room to be themselves as well. Communication cords can also be established between the entity and the affected individual, which ensures a direct line of contact into the individual's psyche.

Where these entities come from is anyone's guess. Are they from another planet within our universe? Maybe. Could they come from another dimension or a parallel universe? That is a possibility as well. In some instances, individuals with alien entities attached to them may have incarnated on the alien's home world sometime in the past. Then, for whatever reason, they choose to migrate and begin incarnating here on earth. Similar to the entities who follow individuals from one lifetime to another, these entities follow from one world or dimension to another.

What an alien entity wants and why they are here is difficult to determine. Many times, they are not interested in our emotional energy. Some attach to us because they want to understand what it is like to live here on earth and have a "human experience." Others come seeking knowledge and life experience, which they glean, from us. Then there are those who, instead of wanting to live vicariously through us, tap into our physical energy. These entities tend to leave us feeling tired, fatigued and drained.

We do not share a common set of values with these entities either. Our life experience here on earth, in most cases, is very different from theirs. What we may view as being right and appropriate here on earth may in their culture be considered abhorrent. Likewise, we may think the same of them, their culture and their traditions.

When the first Spaniards came to the Americas in search of gold, they interacted with the Aztec's who lived in Tenochtitlán, located where Mexico City is today. They were shocked to discover the tradition of human sacrifice, which was practiced by the indigenous population. It was certainly not something a God fearing Christian would do. From their perspective, human sacrifice went against everything they held to be sacred.

Taking this one step further, imagine what it would be like to travel to a foreign world and interact with a race who did not resemble us physically and whose ideas, values, beliefs and rituals were completely different from ours. I am sure it would go far beyond what is depicted in science fiction movies. And no, I have never had an alien entity say to me, "take me to your leader."

The Devil Guy

Christine came to me for a session. She had an entity problem and wanted my assistance to help her move it out. Christine readily interacted with the ghosts she encountered. She admitted that she had the bad habit of trying to heal them and would accidently opening the door to any stray entity who happened to wander by. By the time she came to see me she realized that what she was experiencing was becoming increasingly too much for her to handle.

Christine's aura was loaded with entities. The first thing I saw were eleven entities that hugged the outside of her aura. They seemed to be flattened against the outer wall as if they had been painted on. Then there was the Colonel. The Colonel was a gentleman from the deep south, who had fought for the Confederate army. He had an air of superiority and looked down on women. He was a male chauvinist pig in the truest sense of the word. Last but not least occupying Christine's aura was the "Devil Guy."

The Devil Guy was not from this planet. He appeared in a humanoid form with one exception. Instead of having skin, clothes or facial features he was covered in flames. He reminded me of the Human Torch from the Fantastic 4 comic books, except with an attitude.

We decided to begin work on moving the Devil Guy out during one of our sessions. As I stated earlier, alien entities do not "speak English." This made communicating with him difficult. I did everything in my power to try to understand what he wanted from Christine. All I kept sensing was that he wanted her to go... go with him. He was very insistent. He did not want to wait until tomorrow, "you have to come with me right now." At least that is what I think he was trying to say.

I started to think about his demand. Since Christine was still alive and living here on Earth I could not figure out how she could possibly accommodate him. Did he want her to astral travel? Maybe she was to go with him in her dream state. Then it occurred to me what he really wanted. He wanted her to leave her body, as in permanently. Once dead, she could go with him.

Astonished at his request, I asked Christine if she had ever attempted suicide. Her response, to my surprise, was yes – twice. We were both taken back by the Devil Guy's brashness. How could he ask such a thing? Then again, he was an alien entity.

Demons

The rarest form of all attached alien entities are what I call demons. From a Christian point of view, all attached entities, regardless of their type, are considered demons and casting out demons is acceptable. If a person is being influenced by a dark unseen force, such as Flip Wilson's Geraldine ("the devil made me do it"), then removing its presence from the person's life would be considered an appropriate course of action.

Demons are a very specific group of entities. Demons take the concept of entity attachment one step further and make any other attached entity seem like a simple annoyance. In addition to not having human form, these entities enter the physical body as opposed to only occupying the auric field. This is what makes demons the lowest kind of being on my entity scale.

Demons penetrate the body through its chakras or energy centers. The sufferer may cohabitate the body with the entity or the entity may knock them out of the body all together. If their consciousness stays in the body, the individual will have to deal with the overwhelming effects of the entity in their life. If they are knocked fully out of the body the net result is a full blown possession.

It has been my observation that entities that fall into this category are just plain not nice. They are something right out of the Exorcist and need to be dealt with right away. I have never taken much time to get to know any of the demons I have encountered over the years and I think I will keep it that way.

Sam

I received a call from a man in dire need. He wanted a medical intuition evaluation. Sam complained of losing weight at an alarming rate. In less than a year, he had lost over 75 lbs and now weighed 145 lbs. According to Sam, he was all skin and bones. What worried him the most was the doctors could not figure out what was wrong with him. He believed if his weight kept dropping at its current rate he would die.

Looking to identify what was causing Sam's rapid weight loss the first thing I saw stunned me. Right in the middle of his chest was an alien entity. This "thing" for lack of a better way to describe it, looked like a giant worm or a slug. It had big eyes that glared back at me and it had huge teeth. At first, I was not sure what to say to Sam. I asked him if he was open to discussing his problem from a metaphysical perspective. Thank God he was. Sam had been exploring metaphysical concepts for a very long time. This put me at ease and allowed me to truthfully communicate what I was detecting. Had Sam not been so open to my communication I am not sure what I would have said to him.

This is how our conversation started. "Sam, you have a giant worm-like entity sitting in the middle of your chest."

I paused and waited for a response to see how well my comment was taken. Sam was silent. I continued and explained how I saw the demon sitting right in the middle of his fourth chakra, his heart center. This placement made it impossible for him to make any decisions. This included the decision to eat. I believed he was not mysteriously losing weight because of some unknown disease; it was because he would consistently choose to not eat.

Sam sounded almost relieved when I finished talking. He validated the fact that indeed the biggest part of his problem was that he just could not eat. I explained to him that every time he went into his heart center to make a decision the entity would take over and provide him with answers that were harmful to his wellbeing. For example, when Sam thought he was hungry and wanted to eat, the entity would turn it around and tell him he was full. I would not want to be Sam.

At the time of our session, dealing with this kind of an entity was way out of my league. I recommended to Sam that he consult a demonologist. Demonologists are individuals who specialize in this kind of entity attachment.

Entities and Mental Disorders

According to Wikipedia, "A mental disorder or mental illness is a psychological or behavioral pattern that occurs in an individual and is thought to cause distress or disability that is not expected as part of normal development or culture." Contemporary medicine has identified brain chemistry imbalances, structural issues of the brain or a combination of the two as the cause of most mental health issues. In many instances, their conclusion is correct. The cause of mental illness in some individuals still eludes health care professionals.

When the mental health of an individual is evaluated, their behaviors, psychology, physiology and biochemistry are assessed and a diagnosis is based upon these factors. The notion that someone or something external to the individual is influencing his or her behavior is laughable in the realm of modern medicine. The concept of an entity being attached to an individual is never considered, never seen as a causative factor of a mental illness. To mention the idea of an attached entity to a mental health professional as a contributory factor could put you in the same light as the affected individual.

As we have already discussed, entities can affect our confidence and self-worth as well as take over our thought processes. They can tell us what to do and how to feel. They work diligently to rationalize why what they are saying to us is right, appropriate and preferable. Since they are in our auric field, we assume that their thoughts, justifications, opinions, epiphanies, ideas and insights, however distorted from the truth, are our own. Because of this, we do not question them. Why would we – they are ours.

Our inability to distinguish between our own thoughts and the input we are receiving from the entity is what causes so many problems. This is especially true when we are out of emotional balance. Emotions such as pain, fear or depression keep us from aligning with our spiritual center.

When we are energetically aligned with ourselves, we feel whole, complete, grounded and congruent. From this centered place we can check in with all our parts to see if they are in agreement with one another. We are also able to make decisions from the heart and are not ruled only by our minds, personality, egos or our entities. Being in alignment with our spiritual center can actually help push an entity out of our auric field.

Keeping us out of emotional and energetic balance supports an entity's ability to control us. If we are unable to discern reality clearly, we will accept whatever seems real to us, in that moment, true or not. While this can happen to anyone, attached entity or not, this concept is taken to the extreme when the entity just will not go.

Entities typically move into our auric field when called upon and leave again when we are able to find emotional balance. I believe individuals who suffer from certain mental health issues are unable to get the entity out of their personal space for very long periods of time, if ever. Instead of having moments of interference by the entity in their lives, sufferers have to deal with their presence daily. In this next section we will explore a number of mental health concerns all of which I am convinced involve an entity attachment.

Obsessive Compulsive Disorder

Obsessive Compulsive Disorder (OCD) is an affliction where the sufferer is plagued by repeated and persistent unwanted thoughts, ideas and impulses (the obsession), and the ritual behavior (the compulsion), acted out in an attempt to make the obsession go away. Obsessive thoughts, such as the fear of being contaminated by germs may cause the individual to repeatedly wash their hands. An excessive focus on a concept may cause them to double check things such as locks, appliances or switches. If they are obsessed with a need for order and symmetry in their life, they may arrange things "just so." While the obsessive thoughts may seem senseless to the sufferer often they cannot stop themselves. They are driven by the anxiety they experience when the object of their obsession is not acted upon.

Working with a client with OCD can be challenging at best. Many of them suffer from a sixth chakra (the third eye) on overdrive. Internally, their thoughts jump quickly from one idea to another. As they try communicating their problems the topic of the conversation moves, flows, changes and shifts faster than you can blink an eye. They can become overwhelmed by the sheer number of different thoughts that fly through their heads in any given moment. This makes it difficult for them to focus on a specific topic for very long and causes them to become frustrated when attempting to seek help. Their lack of focus makes headway in any one direction difficult. If they could only get help in realigning their energetic body and restoring health to their chakras, their issues could be minimized if not fully resolved.

This is one form of OCD. There is another kind of OCD that affects individuals that on the outside they may have the same symptoms as the one we have just discussed but energetically they have something more. These individuals have an entity attached to them.

Health care professionals assume it is a biochemical issue affecting the individual that is causing them to obsess. What if it is an entity who is putting the ongoing obsessive thoughts into their heads? Individuals with "attached entity OCD" often have an entity corded to their fifth or sixth chakra. It will remain in the person's auric field for days, weeks, months and even years. If the person is able to temporarily move the entity out of their personal space it will quickly return to pester its host once again with their ongoing and non-stop requests. These individuals can be helped by assisting them in moving the entity out. Entity attachment as the cause of their problems is probably not something their doctors talk to them about, and removing them is unfortunately not an acknowledged treatment protocol.

Denise

One of my clients, Denise, was such an individual. Plagued by ongoing and repetitive thoughts, it was difficult for her to move forward in life. Her entity, Jake, was attached to the side of her head, right by her sixth chakra. He would get her to wash her hands repeatedly to get rid of germs that may have found their way onto her hands. He would have her check household appliances such as the stove or iron over and over again to ensure they were not turned on. Denise did not want to burn down her house did she?

In addition to these ritualistic behaviors, he would also cause her to obsess about other people or subject matter that may have seemed interesting to her (him?) in that moment. It was not unusual for her to sit in front of her computer for hours or even days doing research on a topic that may have caught her eye. She would worry, think and wonder about nearly everything – all of the time.

The impact of the cord going into her sixth chakra also had another affect on her. It kept her in her head. It was extremely difficult for Denise to ground her body especially when all of her energy was tied up in her thoughts. When the physical body is grounded, we are afforded the opportunity to let go of energy that is not ours. It can be likened to a giant bathtub drain which can release unwanted subtle energy from the body. Grounding allows us to rebalance our bodies and helps us to align ourselves with our spiritual center.

Everyone can be affected by not having their bodies grounded. In Denise's case, we had to move Jake out of her auric field before we were able to move forward. Once he was gone Denise was able to ground herself properly. Within a very short period of time her life began to move forward again.

Auditory Hallucinations

Hearing voices is another psychiatric disorder that, in my mind, is directly connected to an attached entity. Everyone at some point in life has talked to themselves. We often have dialogs with ourselves (with our inner voice) about a direction to take, whether to buy something or even what shoes to wear on a night out. If we pay attention to this voice, we would quickly notice that this communication, the voice we are hearing, comes from within our own heads.

Someone who suffers from auditory hallucinations does not hear voices in the same way. In addition to communicating with their own inner voice, they also hear the voice or voices of other individuals. Unfortunately, when observed, there is not anyone within earshot talking to them, or is there?

They say there is a fine line between individuals who have the intuitive ability of clairaudience and those that suffer from auditory hallucinations. People who intentionally communicate with spirit non-verbally also hear voices. Seen as a positive experience, they communicate with angels, spirit guides and even ghosts who remain outside their auric field. These non-corporeal beings respect and honor the individual's boundaries. The intuitive is also able to distinguish between what exists and what does not exist in what is commonly identified as our cognitive reality.

From the outside, it may seem as if the clairaudient individual is suffering from auditory hallucinations and in our current view of the world I guess they are. Let us nevertheless take a step back and look at this concept from an intuitive perspective. Ghosts, angels, spirit guides and entities are around us all the time. We are taught to filter out their presence from our normal waking consciousness. It is not because the beings are not there, it is because we are unconsciously ignoring their existence.

Psychics and other intuitive individuals learn how to "thin the veil" of existence, making what was once unseen and unheard become observable. Like the intuitive who can hear spirit, individuals who experience auditory hallucinations are able to access this unseen world.

It is when the voices they are hearing begin expressing disturbing concepts, are derogatory, command that a task be done or preoccupy our waking existence, that a flag should be raised. If the individual is also unable to distinguish between what is real, and what is coming from the other side, then some kind of action need to be taken.

Individuals who suffer negatively from auditory hallucinations always have one or more entities attached to them. These entities are always of a very low vibration. They tend to hide behind the person's back and will cord them in the back of the fifth or sixth chakras. Once attached, they will talk to them incessantly or provide a running commentary about what is going on. They may even provoke them into performing bizarre or unusual behaviors. Unlike the attached entities who come during times of trouble and then leaves, these entities are in for the long run and rarely leave the sufferer's auric field.

In most cases, these individuals also have vortexes that open into the back of the auric field. This leaves them vulnerable to additional entity attachments. With their "backdoor" open, any entity can walk right in and begin to exert its control over him or her. It is not uncommon to have additional vortexes, both big and small, form in the back of their auric field once a door has been opened. This creates even easier access for the entities to come and go.

Brian

Brian came to me because he was hearing voices. Without much more information than that, I decided to take him as a client. During our first meeting, he explained that he had been hearing multiple voices for the last seven years. Their messages, although annoying to him, were never mean, cruel or spiteful. As time went on all of this changed.

It was his observation that in recent months the voices had become louder and more intrusive. He was able to keep their constant dialog in the background for a long time. Now they had moved into the foreground of his thoughts. He feared everything and worried incessantly. Paranoia invaded every aspect of his life. There were moments, he admitted, that he had thoughts of suicide and times when he could not leave his home for fear of what may happen "out there." What scared him the most were the thoughts of hurting others, he was currently having. This is what brought him to me that day.

I looked at his energy field. I could see several issues within his chakra system right away. A number of his chakras were blown out. They were receiving too much information leaving him unable to filter out the entity's communication. I also detected three entities in his auric field as well as a number of portals leading into his personal space.

We did some work clearing the entities out of his energy field. By the end of the session, I was still very concerned. The work we needed to do to restore Brian's health, although an effective course of action, might take too long. I felt that in the interim he could be a threat to himself and others. I strongly suggested to Brian that he talk to a doctor about medication. While not an advocate of prescription drugs, in this instance it was the most prudent choice. I believed the drugs would help him deal with the entities, although I was not sure how at that time. Thankfully, he listened to my suggestion and got help.

A few weeks later he came back for another session. I was fascinated by the way the medication worked. In our first session Brian's auric field was wide open. He was receiving more information than he wanted. Now it was closed down tighter than a drum. Instead of being able to send, receive and assimilate information from the world around him, something we are all designed to do, his energetic sensors had been turned off, shut down and somehow blocked. This left him, in a way, blind to the energetic world around him. He was unable to experience "reality" in the same way we experience reality. The medication also interfered with our ability to move forward in our healing endeavor. It was impossible to deal with his entity issues while taking the medication and we were both too afraid to have him stop taking them.

Dissociative Disorders

Before we move on and discuss dissociative disorders, I would like to talk about another psychic ability that allows an individual to interact with spirit, namely channeling. Channeling is a technique by which an individual allows an entity to communicate, not to them, but through them. Individuals who channel entities such as spirit guides, angels or ascended masters, are like individuals who communicate with beings utilizing their clairaudience. They choose to open themselves up and work to create this energetic dynamic within themselves. They too are not suffering from a mental disorder.

Channeling is often done with tight controls and a deep understanding of its implications. Individuals who channel spirit are aware of what they are doing and choose to be a conduit for the transmission of spiritual information. This information can be channeled in two different ways. The first, least invasive and safest way is to have the entity remain outside the auric field and communicate via the seventh chakra, the energy center located at the top of the head.

The seventh chakra is the energy center by which we interact with our higher self. It is where we receive inspiration from God. Information received by an individual via the seventh chakra bypasses their egos. Because it bypasses the ego the message received is not interpreted by the conscious mind. By moving the ego out of the way, an undeterred stream of consciousness can flow through them. The message is communicated to the individual either in written form, such as when performing automatic writing, or right out of the channeler's mouth.

The other method employed by channelers is where the individual allows the entity to enter into their physical bodies, either partially or fully. The chance of an entity attaching to us if it is kept out of the auric field is fairly low. Individuals who bring an entity either partially or fully into their bodies are at a greater risk. This is especially true of channelers who perform what is called full body channeling.

Full body channeling implies that an individual allows an entity to come all the way into their bodies. This allows the being to have total and complete control over the physical body. The voice, mannerisms, facial expressions and personality are those of the entity; well at least as well as the host can facilitate it. The host, in turn, steps out of the body giving up his or her control. In most cases, the host has no recall of what transpired when the entity was present.

In this form of channeling there is always an agreement between the channeler and the entity which states that the entity will fully detach from his or her body when the session is over or after the desired information has been received. It is when the individual leaves their bodies and the entity does not go away when the session is completed that problems can occur. Fortunately, instances of this occurring are very rare.

Greg

While living in the Seattle, Washington area, I had the opportunity to do psychic readings out of a metaphysical bookstore. Joining me was another psychic named Greg. Greg was a full body channel. During a typical session, he would bring through one of the two entities he regularly worked with. I had never been around or experienced a channel at work so it was a real thrill to get to see him in action.

I was always fascinated to watch Greg bring his entities in. Depending on the client, one of his entities would enter his body and communicate through him. The first of his entities was Roan. He seemed young and was full of practical advice. When Greg allowed Roan inside his body, he would only come in as far as Greg's chest.

The second entity Greg worked with was Gabriel. Gabriel was old and wise. When it was his turn to enter Greg's body, he, Gabriel, filled it completely. As Gabriel entered Greg's body, it would begin to contort. He would lean over and rest the weight of his body on his left elbow. As Gabriel moved further in, Greg's right arm would begin to contract and he would draw it in close to his body. His back would then twist until his upper body was positioned sideways to the table. His neck would continue bending and twisting even further over to the left. I could sense the tightness and pain in his contorted body.

When the scheduled session was over, as quickly as it had come on, each entity would leave Greg's body effortlessly. Greg was always curious to know how I could tell which entity he had brought in for his sessions. When I described what I saw, especially when he brought Gabriel into his body, he was unaware of any changes he made to his posture. This was because when Gabriel took control, Greg's body would change and shift position reflecting Gabriel's image of himself. As Gabriel exited, Greg's body would return to its full and upright position.

Dissociative Disorders

Everyone has "spaced out" at one time or another. This commonplace event is a form of disassociation. When we space out, we, as spirit, leave our body. Sometimes we stay in close proximity to our bodies and watch events going on in our lives as an observer. At other times we go to work, visit our homes, play with our children or travel back to a previous time in our lives. We can even travel out into space or visit foreign and exotic lands. Wherever our mind and imagination can take us we can go, separate from our bodies.

When someone suffers from a dissociative disorder they chronically try to escape reality often in unhealthy ways. This pattern of behavior usually develops as a defense or coping mechanism where the affected individual emotionally detaches from a situation that is too traumatic to bear. Dissociation can also work to protect us from reliving difficult and painful memories.

Once we step outside of our bodies, the stage is set for an entity to come in. Like Greg's interaction with Roan, the further we step away from our physical bodies, the further in they can enter. If our dissociation is complete, as with Gabriel, they can take complete control over us and our physical bodies, leaving us out in the cold.

There are four different types of dissociative disorders classified by the American Medical Association.

Based upon my observations, I believe that everyone who suffers from any kind of an entity attachment also suffers from some form or severity of a dissociative disorder. The most common is depersonalization disorder. Depersonalization disorder is characterized by a sudden sense of being outside yourself and observing what is going on to you or around you. It is as if you are watching a movie and you are the star. Some people experience it as if they are going through the motions of life but not experiencing it. They might feel disconnected from their body or, what is going on is just a dream.

Probably the best-known form of this disorder is dissociative identity disorder, formally known as multiple personality disorder. In this disorder, the individual switches from one identity to another. This switch typically occurs when the individual is under severe stress. These individuals are plagued by multiple entities who will take over the body at different times. When we talked about entities earlier, we discussed how an entity will come to us based upon an emotional vibration we are sending out. Thus, depending on which emotion is being triggered, a different entity will be called in. I believe there is a direct correlation between the number of entities attached to these individuals and the different types of traumatic events they have experienced.

Someone suffering from dissociative amnesia will experience sudden memory loss after a traumatic event. This is often the case with children who suffer from ongoing and repeated sexual abuse. In situations such as this, they leave their bodies in order to deal with the event. It is often many years after the traumatic event that these individuals are able to recall the nightmare they have experienced.

The final form of this disorder, although rare, is dissociative fugue. Here, a person loses all sense of who they are. They will typically leave work or home abruptly and might try to create a new identity for themselves in a new location. When the individual comes out of the fugue state, that is, when the entity has finally vacated their physical body and energetic space, the individual feels disoriented and has no memory about what happed during the period of their absence.

There are a multitude of other mental health disorders that most likely have entity attachment as their root cause. I look forward to the day when health care looks beyond the physical and considers entity attachment as a potential source of mental illness.

What To Do, What To Do… I Don't Need All Of This Excitement!

Dark Angels

What To Do, What To Do...

You have no power here! Now be gone, before somebody drops a house on you!
  – Glenda, the Good Witch of the North

In the previous sections, we discussed a number of the different "dark angels" that interact with us and our world. So what do you do if you find you have an unwanted visitor in your home or somehow affecting your life? Hoping it will go away might sound like an effective solution but it will probably not fix the problem. As we move forward, we will be exploring a number of techniques you can use to clear your home of residual energy. We will be discussing how you can get rid of pesky ghosts as well as how to move an attached entity out of your personal space. We will also discuss how to help move an entity out of someone's auric field and support it as it transitions into the light.

In the physical world the laying on of hands has historically been used to help restore an individual's state of health and harmony – physically, mentally, emotionally and spiritually. The laying on of hands is a form of energy medicine. With energy medicine, energy IS the medicine. Energy medicine is a term used to describe a variety of alternative and complementary therapies. It is often only associated with healing the physical body. Energy medicine techniques, however, can be used to heal, clear and enliven anything in the physical world and beyond.

According to quantum physicists, everything in the universe is made up of patterns of energy that vibrate at different frequencies. The visible world, the world in which we live, is simply viewed as a denser form or a slower vibration of this energy. This is what makes the things we see and touch seem solid. Like the world around us, we too are made up of energy and patterns of energy. Our vibration matches, or is in synch with, the world in which we live. This allows us to experience each other via our five senses.

Entities, whether an angel, spirit guide, ghost, attached entity or demon vibrate at a rate that is imperceptible to the five senses, thus making them invisible to the naked eye. Since everything in our universe, both seen and unseen, is composed of energy, many energy medicine techniques can be used on a location, an object, a ghost and even an attached entity.

Therapies that fall under the umbrella of energy medicine include acupuncture, qu gong, homeopathy, magnetic, light, sound or crystal therapy, Therapeutic Touch and Reiki. They are all based upon the premise of manipulating and influencing the energetic patterns that make up our being.

Healing, through the use of energy medicine techniques, happens when the movement of subtle energy is restored to the affected parts of the energy field. It is believed that a shift in the flow of our life force (subtle) energy can directly influence our well-being. It raises the vibratory level of the energy field and works to clear the negative thoughts and emotions that can affect the proper functioning of our bodies. Practitioners of energy medicine believe that if we raise our vibratory level; negative energy will break apart and fall away, thus restoring health on all levels.

Of the vast number of healing modalities that make up energy medicine, we will be focusing on the use of energy healing. Energy healing is a subset of energy medicine. It involves the conscious use of the intention by the healer to affect the energy field of the person to be healed.

How can energy healing and energy medicine help? Especially when we are talking about ghosts, attached entities and demons? Simply... In order to get rid of what may be pestering you, steps need to be taken to shift, change, and transmute you, your location and the energy of the unwanted presence in your life. This is what energy healing can and will do.

The Use Of Intention In Healing

When we are able to focus our attention and create a clear intention of the result we desire in a healing session we will dramatically improve our overall success. Here is an example. Rituals are one method of employing energy healing. Rituals create an atmosphere, which accentuate one's intent and intentions for a desired result. The burning of sage or sweet grass (smudging) in the Native American cultures is traditionally used to clear unwanted or negative energies. It is believed that smudging dissipates discordant energy, which is carried out upon the uprising smoke.

Many people complain when performing rituals, like smudging, that they just do not work. They mistakenly believe it is the burning sage alone that clears the unwanted energy. They assume that since they performed the ritual as required, the actions they took should be enough to create a change. This is a false assumption! Waving sage around a person or a room will never work by itself. If however, your intention and your focus is to clear negative energy and as you move the sage's smoke around the affected item you intuitively watch or feel the negativity as it is being lifted, then healing, changing and shifting can occur.

There is a big difference between going through the motions and actually working with energy. One day at a holistic/new age fair, I was performing medical intuition evaluations for attendees. A few rows from where I worked was a practitioner doing energy work. As part of his healing sessions, he would point a crystal towards his client's seventh chakra. The crystal he used was very impressive. It was long and clear and nicely tapered. It was mounted in a wand, which he held between the palms of his hands.

Throughout the day, I would periodically look up and see him in session. On many of these occasions, he would be sitting at the head of his client with wand in hand. Instead of having his energy and intention focused on his client, he would be looking around the room watching people as they walked down the crowded aisles or checking his watch. How much healing was actually happening I could not tell you.

As you will see, rituals, signs, symbols and other tools are not required to produce results on energetic levels. They are wonderful at helping to create a mood and support us in focusing our intention, but we can use our imagination alone to produce the same, if not better, results.

Clearing Stagnant Energy From a Location

Keeping your environment free of stagnant energy can be like a breath of fresh air. This is true for everyone and especially relevant for anyone who believes their house is haunted. Even if they think they might have an entity attached. Stagnant energy in our homes can affect anyone (with or without a body) who enters.

Does your house feel dark, heavy and maybe even a bit depressing? These are all indicators of stagnant energy. Cleaning and clearing your home of trapped energy will not only make the environment feel better but you will personally experience the benefits of the healing as well. It can also help you determine if you have a residual energy problem in your home instead of something else.

Clearing a space of residual energy has been practiced since the earliest of times all over the world. Even today, many religious traditions use incense, sage and water to cleanse their environment. One way to clear your house (or location) is to do a "spring cleaning." Remember spring cleaning? It is when you go through your home and scrub it from top to bottom. Yes, you heard me right. When you do a spring cleaning you clean out your closets, dust the furniture, vacuum the floors and wash the windows. It is also a wonderful time to get rid of things that you are not using which may be carrying residual energy of their own.

We typically only think of doing a spring cleaning when the weather turns nice but it can be done any time of the year. When we do a spring cleaning our intention, our focus, is on cleaning. Our desire is to get rid of the dust, grime and goo that may have accumulated over time. This form of cleansing also works to clear residual energy that may have stagnated in our environment. It happens whether we want it to or not.

If you do not believe me, try it. In the worst-case scenario, you will end up with a spotless home. More likely than not, you will experience a lighter, cleaner, fresher feeling around you. When you are done clearing, take a deep breath. Notice how much more relaxing the space feels. Lightening your home's energetic load, by doing a spring cleaning, is a great way to get the energy in your environment moving and flowing again.

I remember as my kids were growing up we would do an annual spring cleaning. Complaint would abound the entire time we worked. When all was said and done, each of us would remark how much better the house felt. I could never understand why they fought me so much when they always marveled and enjoyed the fruits of our labors.

Smudging

One of the most familiar clearing techniques is smudging. As we already discussed, in Native American traditions the use of burning sage to smudge a person, place or thing has been used for centuries. Dried sage leaf can be purchased in many health food stores very inexpensively. To smudge, light a couple of sage leaves with a match or lighter. Quickly blow out the flame and place the smoldering leaves into a shallow fireproof bowl. Many people like to use a bowl made of abalone for this purpose. Bare in mind, a little goes a long way, so do not go overboard with it or the neighbors might think your house is on fire.

Use your hand or a feather to gently direct the path of the smoke as it rises. Smudge the walls, smudge the furniture, smudge your kids or your spouse. They can only benefit from its healing properties. Move from room to room to give the entire space a healing. Incorporate the power of your intention to create changes and shifts in your home. Hold the intention – "As I clear this space, any stagnant, old or useless energy will be carried out on the rising smoke." Watch as any sluggish, heavy or negative energy begins to dissipate and rise out of the space. Pay attention to any messages you might be receiving. What is your gut telling you? It will often let you know what you need to do. It will also tell you when your work in one area is done.

If sage is not available, or you are in a pinch, the same technique can be done with a stick of your favorite incense. My personal preferences are frankincense and sandalwood. Light a stick of incense and walk around your home using the incense to clear it. As the scented smoke rises into the air, fan it into the areas you want cleared.

You might discover it feels better to move the incense around in certain ways. Perhaps you find yourself compelled to create symbols such as crosses, hearts or stars with the smoke. You could feel guided to say a prayer or work more intensely in one area over another. Regardless of what you are lead to do (and if you feel lead to do something, please follow your inner wisdom), in the end it is not about what you do but about your intention.

Water Purification

Of the many energy healing techniques available, the one I prefer to use when sprucing up the energy in a home is to incorporate water into the healing process. We all recognize the cleaning and restorative properties water provides. By using water I am able fully engage in the clearing process. Being a bit of a clean freak, I like to watch as negative energies are washed out of the room. It increases my inner certainty and amplifies my intentions.

Using water to clear and energize a location does not require many supplies. All you need is some water and something to hold the water. Oh, and do not forget your good intentions.

The most inexpensive method is to use a cup or bowl filled with water. Dip your fingers into the water and spritz it upward into the air. Try using an intention such as – "As I cleanse this space with water, allow any old, negative or stagnant energy to be washed away."

To simplify the process more, I like to use a spray bottle filled with water. Set the spray bottle on mist to make sure none of the items in the room are damaged by the water. Try to use a new bottle or one that has only held water in it. Never use a bottle that been used for household cleaning products.

Crystals or essential oils can be added to charge the water even more. Some people like to say a blessing over the water before they begin. Do whatever feels right to you or what you believe will amplify the healing process. This will support your inner intention and promote the healing and clearing of your home.

When I employ this clearing technique, I like to start in one corner of a room. I begin by spraying water into the upper corner of the room. From there I slowly begin working my way around the room spraying water near the walls in an upward direction. At the same time I imagine a giant drain in the center of the room. I envision the water washing and cleansing the walls and the now dirty water flowing across the floor to the drain and finally out of the room.

Once the outer walls are cleared, I begin making my way into the main part of the room. I like to walk around the room in a spiral that gets smaller and smaller with each revolution. I continue working my way around the room until I find myself in the center of the room standing on the drain I created in my mind's eye. When I feel complete and the task of cleansing the room is done, I will move onto the next room and repeat the process over again. I do this until I have cleared the entire house.

You may detect old, negative or stagnant energy on an item or piece of furniture as you make your way through your home. If you do, give it a spritz or two of water. Watch as the residual energy is carried off to the giant drain in the center of the room. If spraying it with water is not appropriate, clean it, dust it, polish it or shine it with intention. Again, it is not about what you do but about your intention.

When you are done clearing your home, take a few deep breaths and walk from room to room. Notice how good your house feels. It should seem lighter, clearer, happier and less dense.

What To Do, What To Do...

Protecting Your Environment

Once you have cleared your house it is also a good idea to protect it from outside influences. Before we begin, let's start off with a few basics about protecting an environment using energy. To protect a space, be it our homes, apartment or even our cubicle at work, we want to create an energy field around it. This energy field can be likened to our aura. Our aura acts as our personal first line of defense against invaders. It is a part of our subtle energy system and is intrinsic to our make-up as human beings. When we secure our personal space we provide our physical bodies with additional protection. It also helps to create a safe environment for us, one that is free of outside influences.

There are a number of images that can be used to assist us in focusing our attention and supporting our desire to protect our environment. This is one of my favorites. I like to imagine a giant soap bubble surrounding my entire home. I want to be able to see through it and have positive energy and information pass in and out of it easily. This technique can be used if you are in a house, an apartment

Dark Angels

or you share your living space with others. You can even do this around your desk at work to protect yourself from the negative energy of coworkers. Just imagine a giant bubble surrounding YOUR space and not everyone else's.

Why don't you give it a try. Create a protection bubble right now as you are reading along. Take a second and imagine your entire home or location being surrounded by a giant bubble in your mind's eye. As you create it hold the intention – "I am surrounding myself with a protection bubble that will keep me safe and secure." Observe the thin film of the bubble around you. Feel how it creates a barrier between yourself and the area around you. Notice how it allows good, helpful and positive energy in, while keeping negative or harmful energy from entering.

Some people when trying to protect their environment like to use images such as solid brick walls. Images like this do keep everything out and may leave you feeling safer. There is one drawback to this. As they work to keep the bad stuff out, any good, positive or helpful energy is blocked from entering as well.

Another great image you can use to protect an environment is to envision the halo of a candle surrounding you and your space. When I use this image, I feel safe, secure and surrounded by warmth, love and light. It reminds me of sitting in front of a crackling fire at night curled up in a big blanket.

To create a halo of protection, first find a comfortable place to sit. Light a candle in front of you. Look at the golden halo that surrounds the candle. With your eyes open or closed, see in your mind's eye the halo growing. Watch as it gets bigger and bigger and begins to fill your surroundings with its protective energy. Take a deep breath. As you exhale, use your breath to expand the growing light.

Observe it as it expands and fills the entire room. Feel it as it touches the walls around you and then expands even further. Watch it until the entire space is surrounded by its protective glow. Notice how good this feels. Sense the strength and protection your halo of light offers. Ask the halo of protection to keep you and your surroundings safe from any potential negative threats.

Another technique I like to employ is to create an energetic grid of protection. I like to think of a grid of protection as a force field that I have created around my space. You will need a minimum of four crystals to utilize this technique. More can be used if desired, especially if they will help increase your focus, attention and intention. Clear quartz crystals can be used to do this but it really does not matter what kind of crystals you use. The important thing is to select ones you are drawn to or you feel will help to protect and sanctify your living environment.

Place one crystal in each of the four corners of your home. As you put the crystal in place, connect with it. Hold the intention – "This crystal supports the creation of a grid of protection." Imagine the crystal anchoring its energy deeply into the earth. Feel it as this energy penetrates the earth's surface and creates roots deep into the ground. Repeat this process in each of the four corners of the space. If you are in a house and want to include your property as well, you can put the crystals outside on the four corners of the property and create the grid from there. Use as many crystals as you have on hand or feel appropriate to you.

Once all of the crystals are in place, we can begin creating the energy grid. Start in one corner of your location. Imagine the crystal you placed there connecting energetically to the one in the corner to its left. I like to visualize the connection between the crystals as a giant highly charged electrical line or as the intense light of a laser beam. Connect all of the crystals together in your mind's eye until you have created a giant rectangle or square.

Once complete, imagine a line of energy going up the wall until it is the height of your home or surroundings. Do this in each of the four corners. Connect the tops of these energy lines together until when completed you have formed a giant box or cube. I like to create an energetic pyramid over the top of my location as well. This can be done by connecting the upper points of the grid together over the center of your home. As I create this structure, I ask that higher vibrating energy and information to be channeled into my home via the tip of the pyramid.

Now, you have surrounded your home with an energy grid. This protection grid is of your making. As the creator you can control it. Become aware of the movement of the energy in your grid. How does it feel to you? Turn up the power if you do not feel safe or need more protection. Turn it down if it seems too strong or too intense. Decide what kinds of energy you want to keep out and what kinds you do not want to enter. This can include thoughts, beliefs, ghosts, entities and even people. Yes, people too... I am always amazed at how effective working with energy can be in creating change.

One last note about energetic protection. Take the time to evaluate your energetic protection regularly. Regardless if you are using a protection bubble, a halo of protection or a protection grid, they will all need to be tweaked periodically to ensure proper functioning. To do this, imagine your bubble or grid in your mind's eye. Is it still there? Are there holes in it. Is it dark, dirty or decrepit? If it is, repair it, fix it or get rid of it and create a new one. Use this time to set a new intention to the kinds of energies you want coming into your environment. Most of all enjoy the protection and safety it provides.

Working With Ghosts

Not everyone wants to get rid of the ghosts that may be inhabiting their homes. Some people actually believe having a ghost hanging around is fun. I have met many people who enjoy the company their ghosts provide. Instead of being scared of their otherworldly visitors they remember that all ghosts were alive and breathing at one time or another. They recognize that ghosts are people too. They just do not have a body. Each has his or her unique personality, traits, quirks and temperaments. Some are funny and some are tricksters. Others are looking for help, reassurance or guidance.

I know this is going to sound weird and probably very scary to many of you, but if you want to have a relationship with your ghost, communication with it is very important. Instead of running out of the room, talk to it. Say "Hey Boris, what are you up to today?" Try to make friends with it. You can talk to it, ask it for advice or have it watch over your home when you are not around. The more you interact with it the more it will return the favor.

What To Do, What To Do...

You can also think of your resident ghost as one of the family. Individuals who are visited by ghost children, for example, often view it as another child in the household. But it is your home. You must be clear about what you want. Create rules, boundaries and limitations. If, for example, it hides your car keys, let it know you want your keys back right away. Ask it to put the keys back in a place where you can easily find them. Expect that your rules be followed. If not, let it know it will have to go.

Earlier we talked about the Grove in Jefferson, Texas. Mitchell and his wife purchased the Grove because of its haunted reputation. A writer of books on ghosts, they decided there could be no better place to live than in a haunted house. While I was there with Metroplex doing an investigation, I had a video camera plugged into a wall outlet. Out of the blue it just stopped working. Without a second thought, Mitchell let the ghost know what he wanted. In a loud, somewhat frustrated voice he said, "Hey guys, quit it."

I turned the video camera off and right back on. It worked perfectly. I later learned that over the years they have had a lot of problems with their ghosts messing around with the electrical wiring. It was also obvious that Mitchell and his wife were used to their mischievous behavior.

Finding out more about your resident ghost can make you feel more comfortable with having it around. It helps to transform the unknown into the known and can greatly reduce the fear factor. It is like introducing yourself to a stranger who over time becomes a close friend. Do not be afraid. Ask your ghost some questions such as:

- Who are you?
- What is your name?
- Are you male or female?
- How old are you?
- When were you alive?
- Why are you here?
- How did you die?
- Are you associated or related to anyone who is currently living here? If not them, then who?
- Why are you choosing to stay here?
- Do you want to hurt me and/or my family?
- Would you like to pass on into the light and take your next step?

Open yourself up to receiving information on your claircognitive (knowing), clairvoyant (seeing), clairaudient (hearing) or clairsentient (feeling) channels. Go with your first impression and trust what you receive. It may feel as if you are only making it up or it is just your imagination at work. Many times, however, this is how we receive information on intuitive levels.

The easiest channel for most people to tap into is their clairsentient (feeling) channel. We will often experience a sense of warmth or as if we are with an old friend when interacting with friendly ghosts. If you begin to feel anxious, nauseous, scared or have a bad feeling in the pit of your stomach, it might be your own fear or it could be that the ghost in your home is not very nice.

You might receive a visual image of your ghostly visitor in your mind's eye. The pictures we receive are often written off as a product of our overactive imagination. This, nevertheless, IS how we receive information on clairvoyant levels. They say a picture tells a

What To Do, What To Do...

thousand words and much can be determined about a ghost when we have an image in mind. Is your ghost male or female? How old is it? What is it wearing? The style of clothes a ghost is wearing can indicate the period of time or even a location in which it lived.

You just might know the answer to your questions if your knowingness (claircognitive) channel is open. In turn, you might perceive a voice, similar to your inner voice, which lets you know what is going on. This can be an indicator that your hearing (clairaudient) channel is open.

Know when enough is enough, particularly if you determine your ghost is not very friendly or if it is causing disturbances in your home. If you find yourself afraid of it, or if its behavior creates a threat to your safety or the safety of others, then it is time to help it move on.

Clearing A Residual Energy Haunting

As you may recall, a residual energy haunting is comprised of energy which has imprinted on a object or location. There is not, in the truest sense of the word, a real live, living breathing ghost present. The energy residue of a residual haunting will never hurt you but it can leave you feeling uncomfortable or with a very creepy feeling. Thankfully, the energy of a residual energy haunting can be cleared fairly simply.

This may seem like a silly example but it will help you to understand the difference between clearing a room of residual energy verses clearing a residual energy haunting. When we clear a room of residual energy it is like doing dishes. For the most part we can use a sponge, washrag or scrubber and clean the surface of most items we may find in our sink. The energy of a residual energy haunting can more readily be likened to a pot that has food burned onto its surface. To clean it we have to soak them, scrape it and scrub it with steel wool. Sometimes we will have to repeat this process over and over again until it is finally clean.

Many times the residual energy of a ghost is found in specific areas of a home. Techniques for eliminating it are very similar to the ones used to clear stagnant energy from a room. Tools such as sage, incense or water can be used to break-up the energy that is trapped and finally released. Instead of clearing the entire space, focus your attention on any affected areas you may detect. Watch as the energy of the ghost thins and is carried out of the room as the smoke rises. Do this with love in your heart and the ghost's best interest in mind. Continue this process until your effort feels complete.

Do not get angry or frustrated or in a hurry when it seems as if the energy is not clearing fast enough. Sometimes the ghost's energy is so firmly planted in a location it will take a while for it to go. It is not uncommon to have to repeat this process a few times to ensure all of the trapped energy is fully cleared. As it departs, many people experience a shift in the room's energy. They literally are able to detect a change in the feeling of the space, as if something that was there earlier is now mysteriously gone.

You may discover as you try to get rid of what you think is a residual haunting that your ghostly visitor is really stubborn. You may discover that it will leave the area for a while and make you think your work is done only to reappear at a later date. At other times, they just will not budge. Resistance of this kind is often an indicator that there is something else happening. It might not be a residual haunting but an active one.

What To Do, What To Do...

Clearing An Active Haunting

If you discover you are involved in an active haunting and you want it to go, there are a number of techniques you can employ to facilitate the departure. The suggestions provided are just that – suggestions. I am sure there are many other ways of doing the work being described. In my experience, this method has worked time after time. If you find yourself uncertain or afraid, please contact a trained professional to help you. There is nothing worse than doing a bunch of work and have nothing happen or to have things get worse instead of better.

Regardless if they are naughty or nice, the first thing I recommend is asking your ghost its name. Give your ghost a name, something you can call it, if a name is not forthcoming. This will provide you with an anchor for it. For example, the ghost that lived in the basement of my childhood home I refer to as the "Dead Guy." The man who lived in the back yard of my former home I called Tom. I have had clients who wanted to use less than kind words to name their ghosts, especially ones who were creating problems for them. While I can understand why someone would want to do this, it is not a very good tactic, especially if you are dealing with a mean or angry ghost.

When we use derogatory words, or lash out at a ghost with hostility, our words are backed with the energy of anger, fear or frustration. Ghosts can sense this from us. Their response to our hostility would be similar to one we might have. If someone referred to us as a stupid, what would our reaction be? We would probably dig our heels into the sand and not want to do anything the person requested. Emotions such as these, instead of helping to create a solution to the problem, often backfire on us and add fuel to the fire. While your deepest desire might be to have them to go, your interaction with your ghost needs to come from a compassionate place – from a place of love and caring.

Try to always be courteous to ghosts, entities, attached or otherwise. You do not have to like it, but if you want it to leave, you must give it the choice to go on its own. Do not try to forcibly push it out the door. It just will not work. You would not like being treated that way and neither do these disembodied souls. It can only ruffle the ghost's feathers and make the process harder.

With name in hand, ask your ghost some questions such as those suggested earlier. Open yourself up to receiving its reply. As part of your queries, ask if it wants to move on. This question is very important. If they are willing to take their next step it is usually a simple process to facilitate their movement.

What do you do if you do not get any kind of response from your ghost? This may be because your intuitive channels are not open, or you are not paying attention to the messages you are receiving. If you know the lack of response from your ghost is not about you, then you go to "Plan B."

Plan B – Clearing Resistant Ghosts

When a ghost will not take its next step, more often than not, it is because it has some unresolved issues that need to be addressed. The ghost may be filled with pride or apprehensive of what is on the other side. It may have been unable to surrender itself to the wonderful possibilities awaiting. It may be depressed or suffering from a deep emotional trauma. Its ego is clinging onto physical life instead of allowing whatever is going to happen, happen. This causes it to become stuck between worlds. It is not on the physical plane and has not transcended into the higher level of being it truly is.

Helping to heal these emotional wounds is the next level of clearing you can do to help your ghost move on. One of my favorite techniques to facilitate transition is to "ground" them. In electrical terms, objects with an excess charge (either positive or negative) can have the charge removed by a process known as grounding. We have all experienced a discharge of energy when we get shocked after walking across a carpet and then touch something made of metal. The shock is caused by a buildup of static electricity, which was instantaneously discharged as soon as we came into contact with the metal object.

When an item is grounded, the excess charge is transferred from the charged object to the ground. A lightning rod is a classic example of a ground. When the electrically charged lightning hits a lightning rod, the charge, instead of being transferred to the building it is attached to, follows along a wire that is firmly planted within the earth. The wire creates a conducting pathway for the electrical charge which is dissipated deeply within the earth.

Like a lightning rod, we too can create a conducting pathway to discharge excess energy from our bodies. This is called a grounding cord. We talked about communication cords when we discussed attached entities. A communication cord is a line of energy between two individuals. Similarly, a grounding cord is a cord, which anchors us energetically to the planet. It also serves as a vehicle to release stagnant or surplus emotional energy from our bodies. A grounding cord is also a part of our subtle energy system, which includes the auric field and the chakras. Creating a grounding cord is an excellent way to help calm the body, mind and spirit.

As human beings, we all have grounding cords, which anchor ourselves, our physical and subtle bodies into the earth. When someone says they feel grounded, it is because they are enjoying the connection or reconnection with the earth that has been lost at one time or another. Ideally, the line of energy a grounding cord creates should anchor the individual deeply into the earth. For most people this is not the case. In fact, people with dissociative disorder always have grounding issues.

We can create grounding cords for ourselves. If you find yourself afraid of your ghost, it might not be a bad idea to give yourself a grounding cord and allow the fear and worry you are experiencing to flow out of you. The best place to create this energetic connection is at the base of the spine at the first chakra. The first chakra is located between the anus and the genitals and opens downward facing the earth.

Try creating a grounding cord for yourself. Imagine a line of energy, say a redwood tree, is extending out of your first chakra. Watch as it lengthens and moves down deeply into the earth. Have it continue downward until it reaches the center of the planet. Once there, observe it as it creates roots and anchors itself firmly. Create the intention – "I can easily and effortlessly release whatever I am holding onto in my body and psyche." Take a deep breath and just let it all go.

Give this process a little time to work. As your body begins to discharge energy, you might feel as if something is draining out of you or as if the anxiety or tension you were carrying has somehow magically disappeared. After a few moments, check in with your body and notice how it feels. Do you feel lighter? Less stressed? As if a weight

What To Do, What To Do...

were finally taken off your shoulders? Maybe what you were previously experiencing has lessened or gone away completely. This is the power of grounding at work.

One way to help your ghost to discharge trapped emotional energy is to create a grounding cord for the room you are in. The drain you fashioned when clearing out residual energy from your home was in fact a grounding cord. You used it to help discharge the stagnant energy of that location. If your ghost is present in the area you clear, it will also benefit from the effects of being in a grounded environment.

A much more effective way of promoting healing in your ghost is to create a grounding cord specifically for your ghost. Like the healing you experienced as you gave yourself a grounding cord, giving your ghost a grounding cord will provide it an opportunity to discharge any negative or stagnant emotional energy it has not released.

To ground your ghost you will first need to find it. This is actually very simple to do. If you often feel them in your kitchen, go there. If they hang around a bedroom or out in a hallway, make your way to that part of your home. Find somewhere comfortable to sit or bring a chair with you to the intended location. Close your eyes and take a few deep breaths to help quiet your body and your mind. When you are ready, create a new grounding cord for yourself.

Take a few minutes to allow any tension, worry, fear or anxiety to effortlessly flow from you. When you are ready, ask your ghost to sit in a chair next to you. If having the ghost so close makes you feel uncomfortable, ask it to stand in a far corner of the room. Where it is located in proximity to you is not important. You just want to be able to interface with it enough to give it a ground cord.

Whatever you do, do not invite it into your auric field by mistake.

Use your intuitive senses to detect the location of the ghost. With a little effort and practice, it will not take long until you are able to sense its presence in the room – that is, if you have not already. The ghost may try to be tricky and pull its energy back leaving its presence undetectable. Be strong but kind as you request it to make its presence known. I will have to say that there is a certain amount of assertiveness and lots of intuition required when working with an active haunting. Even more is required when dealing with an attached entity. If you are too scared, or do not feel as if you can be the top dog, then please do call for help.

Once located, try communicating with it again. Ask it questions. Open yourself up to seeing what it looks like. Do not worry if you do not get a response or if the image you see is not totally clear. We are only trying to determine where the ghost, and its first chakra, is located. The rest is unimportant.

Use your intention to create a grounding cord for the ghost. That is right. In your mind's eye construct a line of energy that extends from your ghost's first chakra down into the center of the planet. Once created, set the grounding cords intention – "This grounding cord is releasing any energy, emotional or otherwise, that is keeping the ghost trapped here on earth." You do not need to know the who, what, when, where or why it is stuck. Allow it to let go of what it wants or needs to in that moment. Watch as any stagnant or trapped emotional energy drains from its body.

Check in with your ghost periodically over the next few minutes and ask if he or she feels complete. If an answer in words does not come to you, check in on yourself. How do you feel? Do you feel more peaceful and grounded? Have you detected a positive shift in the energy the ghost is projecting? If the answer is yes, try smudging it with sage or burning incense. Ask it to move into the light or take its next step. Watch to see if it makes its transition.

Here is another method of helping your ghost transition. Once it is nice and grounded, imagine it in a giant bubble, like the kind Glenda, the Good Witch of the North used for travel in the *Wizard of Oz*. Watch as it moves up out of your house and travels across the sky and into the light. As it departs, give it your blessing or wish it well.

From time to time providing your ghost with a grounding cord is all the help it needs. At other times it might take a while to promote a deep enough level of healing for it to depart. With some ghosts a sense of trust needs to develop between it and you before it will let go and take its next step. By giving your ghost a grounding cord daily, or as often as you detect it, even the most stubborn and resistant of us will change.

Jethro

A number of years ago, I obtained a job as the Engineering Services Manager at a manufacturing company. Part of my responsibilities was to ensure the incorporation of changes to the companies manufacturing documents. I had worked in this field for over 15 years and was well versed in the procedures necessary to fulfill the required tasks.

One of the engineering managers I regularly interacted with was Jethro. Condescending was his normal tone of voice and demanding was his typical way of getting what he wanted done. It did not take long before I was sick and tired of how I was being treated by him. The way I dealt with him was very interesting and can be applied to stubborn people and the most resistant ghosts. With him, instead of fighting fire with fire, I fought fire with "woo-woo."

I began the practice of grounding Jethro. If I had to go to a meeting and he would be attending, I would give the meeting room a nice big grounding cord. It seemed to work because the shouting matches between Jethro and I, which had been a regular part of the meetings, stopped. This practice quickly grew. Instead of grounding the room, I started giving Jethro a grounding cord. It did not take long before I was giving him a new grounding cord every day.

This literally went on for three months. And then it happened...

One day, out of the blue, Jethro came to my desk and asked me a question. In fact, in addition to the question he also wanted to chitchat with me! I really did not know how to respond and I probably had a very confused look on my face. At the end of our little "talk", he said "thank you" and left. I have to tell you, my office mate and I looked at each other with our jaws hanging wide open. We wanted to check his temperature to see if he was ill. Instead, we shrugged, smiled and left well enough alone.

I found another job shortly after that memorable day and took my next step. I was not able to heal the whole company but whenever Jethro interacted with me after that time, his attitude was one of respect. Thank God for grounding cords.

Detecting And Eliminating An Entity From Your Auric Field

The first step in self-diagnosing an entity in your auric field is to look for behavioral problems, life changes or other indicators listed below.

- A history of physical, emotional or sexual abuse
- Disassociation or being ungrounded
- Memory problems
- Hearing voices or an inner voice that constantly criticizes you
- Repeating patterns of behaviors
- Anxiety or panic attacks
- Irrational bouts of fear, anger, sadness or guilt
- Sudden changes in behavior or mood swings
- Depression or thoughts of suicide that you cannot seem to stop
- Addictive behaviors, including addiction to alcohol, drugs, cigarettes, sex or gambling
- Impulsive behavior or an attraction to dangerous situations

- Illnesses that will not respond to treatment or are of an unknown cause

Take a moment to evaluate your past behaviors. Have there been repeated situations in which you suddenly found yourself experiencing anxiety, fear, anger, depression or any of the behaviors listed? If you have, perhaps it is time to do a little investigative work into what is happening to you. The unexplained change you are undergoing may actually be the entity working its magic in your auric field.

To begin your self-analysis you need to pay attention on your aura. Do you feel some heat or tingling in and around your body. Do you detect it in one location verses all over your body? Do you feel a subtle pressure on your skin or as if someone is standing extremely close to you? These are all indicators that somebody or something is in your personal space. If you do not sense it right away, give it some time. When your entity is in your aura, you will typically sense the same kind and quality of energy, in the same location of your body whenever it is around.

The first step to eliminating an entity from your auric field is to get it out of your personal space. If you detect an entity in your field ask it to move out. In your mind's eye, show it where you want it to go. Personally, I like to have it sit in a chair next to me. Sometimes I have it stand in the corner of the room. Ghosts and entities alike are highly sensitive to our unspoken words, the images we project in our minds eye and also our emotions. When you address it be firm. Be confident! At the same time, do not be mean, spiteful or angry. Just show it where you want it to go and imagine it there. Do not give it any options.

What To Do, What To Do...

As soon as it is out of your auric field give yourself a grounding cord. Create a brand new one that goes from the base of your spine down to the center of the planet. When it is in place, allow it to release any trapped or stagnant energy from your body. Next, evaluate your auric field. Imagine you can fill in any holes or weak spots through which the entity can enter. Envision the walls of your field as a thick or impenetrable soap bubble or like a sheet of Plexiglas.

When you are done, check again to see if your entity is still sitting where you told it to sit. Lay down the law. Tell it that your personal space is yours and you do not want to share it. Let it know if it wants to come around you, it has to stay out of your aura.

Take a deep breath and notice how you feel. Does your body feel more grounded? Are you more relaxed or centered? How do you feel about your thought processes and other symptoms that led you to believe you have an entity? Have they calmed down? Sometimes, especially if we are fully engaged with the entity, moving it out of our space can leave us feeling tired or drained. This is normal. If you find yourself feeling this way, do not worry. Your body is releasing energy and that is a good thing.

By acknowledging how YOU feel in your space, you can begin to recognize what it is like to have you and only you in it. It affords you the opportunity to identify who you are and how your energy feels without the influence of your entity. If you do not feel this way, then take it as a reminder that something else might be going on.

With your entity out of your personal space it is time to try giving it a grounding cord. Over time, your entity may choose to leave on its own.

There are many times when your entity will be in your

space and you will not know it. This is especially true when obsessive thoughts, altered moods, depression, addictive behaviors surface. When we feel depressed or if we are trapped in the never-ending cycle of obsessive thoughts, it is difficult for us to look at what is going on in our space and to recognize it might not be "Me." It is at these times that working with a friend or a trained professional are especially helpful. We will explore why this is true as we revisit my relationship with George.

George –Revisited

You first met George when we discussed attached entities. George was my own personal and up-close entity attachment. Over the years, I experienced numerous occasions when George would enter my auric field and leave me in such a bad place that I was unable to see there was something wrong. I would be trapped in my own negative pity party and the thought that I could help myself never crossed my mind.

This may sound strange since much of the work I do is as a healer, but when you are in "that place", it is hard to get out of your own way. Having a friend or a support network to work with can really help. They can help you recognize when your entity is in your space. They can also remind you there are things you can do to help yourself.

I first learned about George when a close friend of mine, an excellent psychic, named Kim told me about him. I was in one of my moods. The world looked dark and everything in it was hard. Kim first noticed him on one of the many days I found myself stuck in victim energy, an energy George provoked. Without any knowledge of what we were doing and without any methodology to do an entity release, we managed to move him out of my auric field. It felt great to be me again. I think we were both shocked at the dramatic difference it made to my mood. We made an agreement that day that if she ever detected George in my space again she would let me know. She also volunteered to assist me, if needed, to move him out.

It was not long before George was back. I did not realize he was there and up to his old tricks again. But when Kim brought it to my attention, I knew she was right. I asked George to sit in a chair across the room. This took a while to accomplish and required some help from Kim. When all was said and done, I was amazed at my own personal transformation.

With practice, I learned how to move him out of my space by myself. That is, if I could tell he was there in the first place. If I was unsure if George was making an appearance or if I was just having a bad day I would give Kim a quick call. I would ask, "Is George here again?" Her reply always seemed to be yes.

It took about a year to get to the point where George was not constantly bothering me. By that time, two things had happened. 1) I learned what it felt like to have George's energy in my space. Once detected, I could easily move him out. 2) I had been learning and growing, thus creating a place within myself where he was no longer needed. It was not that George had left. I simply was not constantly calling him back into my auric field.

Our reaction to stressful situations change as we take back our autonomy and learn to love ourselves. Instead of going into the dysfunctional thought patterns that keep the entity tied to us, we create new ways of dealing and responding to life's challenges. We no longer, or less often, get on our Bat phone and call our entity back.

It was many years later when I realized that George

was still around. The days of overwhelming doom and gloom had thankfully faded into the distant past but I was becoming increasing aware of other aspects of my emotional self needing attention. I became conscious of a series of negative tapes that would begin playing when certain things would happen in my life. I had not considered these tapes as associated with George but they were.

It was finally only in recent years that I finally moved George out of my life completely. It was by a stroke of sheer luck how this took place. I scheduled a hypnotherapy session with my friend, Carol Layman of Journey Between Lives, to try to clear up a long-standing sinus issue. I had in the past, referred a number of clients to her who were in need of healing from their spirit guides. At the beginning of the session I suggested that if we had time we might try and figure out where George had come from, what he wanted and how I could get rid of him.

What came up during our session was astonishing. This is what I saw. I was a woman who lived in a small town in 15th or 16th century Europe. I was married to George in that lifetime. George was a very controlling man who dominated my every thought and action. As my history unfolded, I was hurrying home from a doctor's appointment. I was rushing home because if George knew I had left the house without his knowledge, and permission, I would be in a lot of trouble. I had decided not to tell him about the doctor's appointment because I thought I might be pregnant. I knew deep down if my suspicions were true I would have to warm him up to the idea of the pregnancy or potentially face his wrath.

Walking as fast as I could along the dirt streets that went through the town, I arrived home only to find

What To Do, What To Do...

George already there. In a rage, he demanded to know where I was. I was afraid to drop the bombshell and admit the suspected pregnancy to him. I tried to avoid the issue and told him that I had gone into town to run a few errands. That was not enough for George. Overly possessive and jealous, he wanted to know in no uncertain terms who I was seeing. He was convinced that I had gone to see another man and was having a secret rendezvous with him. My assurances that nothing happened only increased the fury in his voice and the insanity in his eyes.

It was not long before George slapped me hard across the face. I fell to the ground. Crying now, George again demanded to know where I went. I pleaded with him to believe me. He did not. With each tear I shed his jealousy and anger grew. Grabbing me by the shoulders, he began to shake my whole body violently. Then with the last bit of his hostility, he grabbed me by the hair and smashed my head it into a piece of furniture several times. My life with George ended in that moment.

I was astounded by how fearful I was of George. Filled with shame, I accepted his repeated abuse. With nowhere else to go, I remained a silent victim. I had no self-esteem. George had seen to that. I was nobody. He brainwashed me into believing that the only person who would love me and take care of me was him. I was also lead to believe that I was a worthless piece of trash. And when he was in my personal space in this lifetime that is exactly how I felt.

Well I have to tell you, the transformation only took seconds to accomplish. Looking George straight in the eyes I snickered at him and told him directly "I don't think so." I finally understood what the game was, and how it was being played, and each time it was at my expense. It did not take a rocket scientist to figure it out. I said "no" to the game George was playing and meant it.

Since that day, I have gratefully never experienced George again in my life. Where he went, I do not know. What I do know is that he is not around me anymore. My life changed after that session. When things would happen that would ordinarily start the negative tapes rolling, they just were not there. I remember on one occasion feeling a bit odd when the tapes I had become so accustomed to, did not start. It was a weird sensation but a freeing one. Goodbye George!

What To Do, What To Do...

Helping Someone Get Rid Of An Attached Entity For Good

Hypnotherapy is one method of permanently ridding yourself of an attached entity. When I am working with clients, I take a different approach. This approach relies on intuition and the ability to work with the entity's subtle energy.

The steps I will be presenting should be done with caution. Trying to permanently remove an attached entity is not something you should do on your own. You are always welcome to try but if you have any doubts in your skills and abilities, I would err on the side of caution. There are a number of reasons for this. Entities are interested in staying, some more than others. Many times when an entity is in our auric field we are unable to think straight much less entertain thoughts on how to move it out. They will try to keep us focused on anything and everything else.

You also do not want to accidentally draw more negative entities to you. If you have a really nasty and deeply entrenched entity, rattling its cage is probably not a good idea either. In the best-case scenario, when working by yourself, it will leave. In a worst-case scenario, it will get upset. In the end, it will continue affecting you.

Working with a friend or trained professional has a number of benefits that can be critical to your success. Bear in mind, an outsider is not being affected by the entity. This affords him or her, the opportunity to stay calm, clear and most importantly neutral, when dealing with an entity attachment. Working with an entity can be like a huge power struggle. If you are angry or frustrated it will react in kind. Patience, love and a bit of psychology make the process of moving an entity out easier and more effective.

When endeavoring to have an entity leave for good, it must also be understood that both parties, this means both you and your entity, must be willing to make the transition. We often do not think about it but we also have an investment in the relationship. We are the ones who let the entity into our space in the first place. At that point the relationship benefitted us.

The concept of holding on to something that is harmful has been seen in the individual who is physically ill. They receive a benefit, however distorted, from their illness. Benefits can include sympathy or attention from others. Their illness may keep them from doing things they really do not want to do. It can help them avoid unpleasant issues or personal problems or it may enable them to have the free time they need or have desired.

To release an entity we must be willing to let it go fully. We must also be prepared for any changes that may appear in our lives. What they (the entities) do not want you to know is that you have the power to move them out. And like the ruby slippers of Oz, you have always had that power within you. You are the one with the body – not the

What To Do, What To Do...

entity.

We have already talked about why some ghosts or entities choose to stay. But what about you? Do you still have some unresolved issues that your entity is still supporting? How will your life be affected if the entity did leave? If the originating issues have not been identified, addressed or resolved, then the entity in question might not move on. For example, if the entity originally attached because of a drinking problem, then you will need to abstain from alcohol before a successful removal can occur.

As we continue, please note, as opposed to describing work you can do on yourself, the methods and techniques described should be done on a friend, by a friend or by trained professional. When referring to you, my intention is to describe you as the friend or professional, performing the work. In turn, the affected individual is referred to as the client or "she."

Our ultimate goal in this section is to have the entity detach, go into the light and take its next step. On occasion it will go on its own, especially after it receives a healing. In most instances, you will have to finagle, cajole or even call in deceased friends or family members of the entity to help. In one client's case, a "ghost dog" came into the session and helped the entity take its next step. The dog was his trusted companion as a child and he gladly went with his long time friend into the light.

Providing a precise methodology for eliminating an attached entity from someone's life is difficult to do. It all depends on the personality of the entity. What an entity wants and needs in order to take its next step can vary greatly. As we continue, I will provide a guideline you can follow. The best way to deal with what may come up during a session is to go with your gut feeling and allow your intuition to be your guide.

Start the session by giving yourself and the room you are in, grounding cords. Set the intention that any trapped or negative energy will be released, thus creating a neutral environment in which to work. Take a few deep breaths and when you are ready invite your friend, your client into the room and begin.

When communicating with an entity it is best to have your client talk directly to the entity. To help facilitate this communication, have your client close her eyes and take a few deep breaths. Give your client a grounding cord to support her in the work the two of you are about to begin. Suggest to your client to relax and open herself up to receiving information from her entity. Let her know that the insights may come in the form of a feeling, an image in her minds eye, a knowingness or even a verbal communication. Assure her that any of these ways of receiving information is perfectly normal and should not be dismissed.

Once your client is comfortable, have her open her eyes. Ask your client if she feels some heat or tingling in one area of her body, or if she detects a subtle pressure on her skin. This is an indication of the entity in her personal space. Have her ask the entity to move out of her space and sit in a chair of your choosing. Suggest to her that she be firm, yet loving with her request. Ask your client if she can detect the presence of the entity in the chair? Invite her to give her first impression of the entity. Inquire of her, "In your mind's eye, can you see what it looks like?" Have her check in with her emotions and ask, "How does the presence of the entity make you feel?"

Have your client ask the entity for its name. Tell your client to report the first thing that comes to her mind. If a name is not forthcoming, give the entity a name. Then spend some time asking it questions such as:

- Are you male or female?
- How old are you?
- When were you alive?
- Why are you here?
- How did you die?
- Why did he or she attach to your client?
- What attracted him or her to your client in the first place?
- When did it enter his or her auric field the first time?
- How long was it intending to stay?
- Why is it choosing to stay?
- Do you want to hurt "me" or "my family"?
- Would you like to move into the light and take your next step?

You can ask the entity anything you want. The sky is the limit.

Pay attention to insights you may also receive as you ask questions of your client. Not everyone is able to access his or her intuitive information. This might necessitate you to communicate with the entity directly. Give your client the opportunity to answer your question but if she falters, take it upon yourself to act on her behalf. As you ask your client the questions listed above, what answers do you receive? What is the entity saying to you? When your client is done providing any insights she may have received, communicate any thoughts or ideas you may have gleaned.

Give the entity a grounding cord when you are satisfied with your questioning. Watch as the energy that has kept it trapped here on earth drains from its body. Acknowledge any fears or misgivings it may be experiencing. Absolve it of its guilt and let it know it is forgiven. Ask it if there is anything else it needs to release, and using your intention, have it be so.

It is typically at this point that if there is going to be a problem it will rear its ugly head. It may take a few sessions to help the entity fully release its issues and move on. Do not get frustrated. It is just part of the work.

When you feel this process is complete, using your intention, ask for a vortex to "the other side" to appear in the corner of the room. Watch as a beam of light appears and a hole in the fabric of space begins to form. Ask the entity to step into the light. If the entity is hesitant or resistant, ask for a loved one of the entity to step into the room. Observe the requested individual as he or she steps forward. Direct the entity's attention to him or her. Suggest to the attached entity that it can go with its loved one into the light. Watch as they move together into the vortex of light. Observe the vortex as it closes behind them.

A shift in the room's energy is always experienced when an entity has moved on and the vortex closes. Ask your client if she could feel the shift too. Have your client take a few deep breaths and relax. You might want to take a few deep breaths yourself. The work of releasing the entity is done.

Fred – Revisited

We first met Amy when we discussed the concept of incubi and sucubi in the last section. After our initial consultation, Amy decided to come see me again to begin work on eliminating her entity Fred from her life. We

What To Do, What To Do…

started with a brief relaxation exercise. I asked Amy if she could feel Fred's presence in her personal space. I pointed out the general area in her auric field where I detected him. To her surprise, she could feel him pressed right up against her body. I asked her to direct him to sit in the extra chair in my office. It took a while but after a bit of coaxing he finally sat where he was directed.

"What is his name?" I asked. "Fred" she said very quickly. I tried to create a dialog between Amy, Fred and myself. Fred was unresponsive. We never learned much about what he wanted, where he came from and why he was there. This was disappointing but we continued.

Next, we began the process of helping him move on. I suggested to him that he could go into the light. He did NOT want to go. I tried a few different techniques to help him move on but nothing seemed to work. I thought maybe, just maybe he would be more inclined to take his next step if I call in a deceased family member.

Fred's mother stepped forward right away. As she moved into his awareness, he became very upset, or should I say, more upset than he already was. Apparently having his mother come to his aid was a big mistake. He really did not want to see her and began to shut down even more. The level of fear and trauma I sensed around him was overwhelming.

I was fascinated by his intense emotional outburst. It was the same feeling I detected when I encountered the fear/trauma energy in Amy's space. I let Amy know that either this entity was creating both problems (the feelings of extreme fear as well as the overwhelming sexual energy around her) or she actually had two entities attached to her.

I asked Fred's mother to take a step back. I again asked Fred if he would go, but he still would not budge. I decided to begin doing energy work on him. I gave him a grounding cord, cleared his chakras and his auric field. I could feel him start to calm down. His emotions had returned to the level of fear and terror he was accustomed to having.

Then suddenly something unexpected happened. Off to Fred's left, a male figure appeared and moved forward. Walking crouched over, as if trying to avoid walking through the light of a projector at the movie theater, he held out his hand to Fred. It was his older brother. They had always been close in life. At first, Fred did not want to acknowledge his presence. But as his brother drew closer they starting talking to each other – if that is what you would call it.

His brother began showing Fred images of what it was like on the other side. He also shared the feeling, the emotional vibration of this new place. The more he communicated with Fred, the calmer Fred became. The calmer Fred became, the more the light, the opening to the other side, broadened. After about three or four minutes of this exchange Fred took his brother's hand, stood up and started walking into the light.

It was a remarkable experience. I could see an assemblage of friends and family members waiting for Fred on the other side. It reminded me of the scene from the movie Ghost, with Demi Moore, Patrick Swayze and Whoopie Goldburg, where Patrick Swasie is greeted by a whole group of beings on the other side.

Once Fred moved fully in the light the vortex closed behind Fred. I could feel a distinct shift in the energy of the room. Amy could also tell that something, "something magical" had just happened. She also reported feeling better, lighter – as if she had finally rid herself of something that had been weighing her down for years. Had I not been there and experienced this miracle, I do not know if I could believe it was true – but it was.

That's All, Folks…

> *I don't believe that ghosts are "spirits of the dead" because I don't believe in death. In the multiverse, once you're possible, you exist. And once you exist, you exist forever one way or another. Besides, death is the absence of life, and the ghosts I've met are very much alive. What we call ghosts are lifeforms just as you and I are.*
> – Paul F. Eno, Footsteps in the Attic

Ghosts, angels, spirit guides and entities are around us all the time. As human beings, we are taught to filter out their presence from our normal waking consciousness. We are told as children they do not exist. If we do come across one, and share our story with others, we are often told it is our imagination at work or we are looked at like we are crazy. The truth is; ghost, spirits and entities are real. Their impact on us and our lives is unmistakable.

Capturing a "ghost" on film, or with other ghost hunting devices, is rare. This is unfortunate. Many people are waiting for the smoking gun or some form of hard evidence before they will believe in the existence of these ethereal beings. And even if they had the perfect image of an apparition in hand or an EVP recording (electronic voice phenomena) of one of them saying, "I am the ghost of your Aunt Sally," the skeptic will most likely try to debunk its authenticity.

Fortunately, non-corporeal beings can be readily detected and observed on intuitive levels. For the beginner, not trusting what one is seeing, feeling and hearing is often the biggest obstacle to fully believing in oneself. Trust in ourselves builds through time and experience. Trust forms when others validate our experiences. When the people around us, such as a homeowner, also detect what we are sensing then our impressions are being validated. Through this, we learn to believe in what we are picking up. It transforms what we may at first think of as a bizarre coincidence into something real.

As you begin to explore the world of ghosts, spirits and attached entities, do not be afraid. Instead, open yourself intuitively to their presence. Ask them questions. Feel into their emotional energy. Slip your "ghost glasses" on and take a quick peek. As you find out more about the beings you encounter you will discover that they typically are not at all bad or evil. In most cases, they are scared and too frightened to take their next step. Find compassion for these lost souls. Your kindness and understanding will help to set them free.

Home Protection Kits

What do you do if you find you have an unwanted ghostly visitor in your home perhaps affecting your life? Hoping it will go away might sound like an effective solution but it will probably not fix the problem. Our Home Protection Kits can help you to get rid of a pesky ghost. Our Home Protection Kits can raise the vibratory level of your home and work to clear the negative energies that may be affecting you and your environment.

How can our Home Protection Kits help? Especially when we are talking about ghosts? Simply... In order to get rid of what may be pestering you, steps need to be taken to shift, change, and transmute you, your location and the energy of the unwanted presence in your life.

When a ghost will not take its next step, more often than not, it is because it has some unresolved issues that need to be addressed. The ghost may be filled with pride or apprehensive of what is on the other side. It may be unable to surrender itself to the wonderful possibilities awaiting. It may be depressed or suffering from a deep emotional trauma. Its ego is clinging onto physical life instead of allowing whatever is going to happen, happen. Helping to heal these emotional wounds can help your ghost move on.

Our Home Protection Kits can also be used to create an energetic grid for healing and protection around an environment. They are designed to protect a space, be it our homes, apartment or even your cubicle at work. They can be used to create a safe place for you to be and work to keep your location free of outside influences.

Each Home Protection Kit comes with 4 specially selected healing stones. Each healing stone has been cleared of negative energy and is charged with the energy and vibration of its intended use. All Home Protection Kits also include a velour pouch and a step-by-step instruction guide for their use.

Clear Negative Energy

Our Clear Negative Energy kit will help you dispel harmful energies. This kit can be used to shield you and your environment from unwanted influences. This kit is designed to protect you from the negative effects of friends, family, neighbors or any ghostly presence that may be affecting you. This kit also guards against psychic attack. The Clear Negative Energy Kit heals energy blockages, especially for those who are unable or unwilling to let go of the past. Its healing ability can help to eliminate negative emotions such as anger, jealousy, fear and resentment.

Transmute Lower Vibrations

Experience an environment filled with wonderful peaceful energy. Our Transmute Lower Vibrations Kit does just that. This powerful and protective kit transforms lower vibrations into higher ones. It can be used to change the negative energy you may be experiencing in your home. The natural calming properties and strong healing powers of this kit can help to relieve stress and sooth irritability. It can be used to dissolve feelings of sadness and grief while encouraging and promoting our own inner strength. It can also be used to dispel feelings of rage, anger, fear and anxiety and support a state of peace and tranquility.

Amplify Energy & Intention

Our Amplify Energy & Intention Kit supports all of your healing and protection work. It is designed to be used alone or in conjunction with one of our other Home Protection Kits. The Amplify Energy and Intention kit allows you to set your own healing intention and works to augment the power of your intentions. This kit is very programmable and will hold your intention for extended periods of time. It can also be used with one of our other Home Protection Kits to amplify their healing and protective qualities.

To order your home protection kit visit:
www.SoulHealer.com

About Dr. Rita Louise PhD, ND

Bestselling author Dr. Rita Louise is the founder of the Institute of Applied Energetics and the host of Just Energy Radio. She is a Naturopathic Physician and a 20-year veteran in the Human Potential Field. Her unique gift as a medical intuitive and clairvoyant illuminates and enlivens her work.

Author of the books *Man-Made: The Chronicles Of Our Extraterrestrial Gods*, *Avoiding The Cosmic 2X4*, *Dark Angels: An Insider's Guide To Ghosts, Spirits & Attached Entities* and *The Power Within*, Dr. Louise credits early childhood influences for the direction her life would take.

By the age of 8, she began searching for spiritual self-discovery pursuing topics including health and wellness, philosophy and the esoteric arts and sciences, including a deep interest in archeology and human origins. At 30, Dr. Louise enrolled at the Berkeley Psychic Institute where she studied meditation, energy medicine, and learned how to perform clairvoyant readings. Continuing her educational pursuits, Dr. Louise earned the distinction of Reiki Master and Certified Hypnotherapist. At 37, Dr. Louise returned to school full time, earning a degree as a Naturopath and then her Ph.D. in Natural Health Counseling, thus rounding out her education.

A frequent consultant to the media, Dr. Louise has appeared on television and has been a featured guest on many radio shows such as *Coast to Coast w/George Noory*, the *Jerry Pippin X-Zone Radio,* 21st Century Radio, The Kevin Smith Show, , *Contact Radio*, *Second Site Radio*, and *Out Of This World Radio*. She also has been heard monthly on *Feet-To-The-Fire Radio* with her "*SoulHealer Moments*".

Dr. Louise has appeared as a keynote speaker at events such as the *Paradigm Symposium*, *Whole Life Expo*, *ASPE Paranormal Symposium* and the *Texas Ghost Show*. She has also spoken to organizations such as Naturopathic Medical Association, MUFON, the Institute Of Noetic Sciences, and the Texas Department of Health.

She is also a regular contributor to publications including *New Dawn Magazine*, *Atlantis Rising*, *In Light Times* and The Edge News. Her alternative health articles have also appeared in publications such as Fate Magazine, Today's Dallas Woman, Inner Self Magazine, Holistic Health News and The Psychic Journal.

The founder of the North End Psychic Fair and a former Pastor for the Church of Divine Man, Dr. Louise currently trains students in medical intuition, intuitive counseling and energy medicine. In addition, she owns a private practice and is a professor of Alternative Health Studies at Westbrook University.

She is the Chairman of the Board for the International Association of Medical Intuitives and has served on the Board of Directors for the Holistic Chamber of Commerce. Through her work, Dr. Louise has been recognized by the National Register's Who's Who in Executives and Professions.

Other Works by Dr. Rita Louise

Man-Made: The Chronicles Of Our Extraterrestrial Gods

Brace yourself for an astonishing new journey through time! This journey encourages you to take a fresh look at the Gods of myths and legends as they appear in our biblical, oral and written histories. This journey goes back to the beginning of time and recreates history based upon the accounts of the Gods as provided by ancient civilizations from around the world. What surfaces when these primitive sources are put together into a single time continuum is a distinct, yet untold, account of our past - one we were not taught in school.

Man-Made: The Chronicles Of Our Extraterrestrial Gods focuses on many of our most ancient stories. Starting with, "In the Beginning", and ending with the development of civilization as we know it. *Man-Made* tracks the movement of the Gods during a time when they walked the Earth. It brings together the myths from many cultures including the Sumerians, the Greeks, the Maya and the Aborigines of Australia. Each story contains a piece of evidence, a breadcrumb or specific detail that collectively fills in many particulars of our past and provides a bigger broader picture of ourselves. This permits us to unravel contemporary thought and provides a more complete and comprehensive understanding of our past.

Man-Made also poses a novel thought to the reader. What if our ancestors, the Gods of myth, the Gods that were written about, sung about and praised...were not from this world? What if they were not human at all? What if the assumptions made by mainstream archaeologists and historians are wrong and the bizarre stories told by our ancestors are reliably true tales? The reader is asked to take another look at the myths and legends of old. Except, this time we replace the word God with non-terrestrial, non-human or the actual names provided by our ancestors. This slight change provides us an opportunity to look at our history in a whole new way.

""Man-Made" is a flight throughout our mythology and legends, inviting the reader to look at it with new, wide-open eyes."
Philip Coppens
Author of The Ancient Alien Question

Man-Made" is a must-read for all students of myth and global creation stories.
Barbara Hand Clow
Author of Awakening the Planetary Mind

Avoiding The Cosmic 2x4

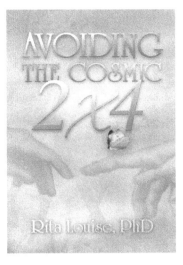

After 15 years of research into the constitution of our subtle energy, Dr. Rita Louise proposes a comprehensive explication of the energetic principles that underscore health and disease.

Avoiding The Cosmic 2x4 is a journey deep into the realm of subtle energy, where it explores the elusive structures that lay hidden to the naked eye but are an integral part of who we are. Through an integration of the Yogic chakra system and the Qabalah, the basis of Jewish mysticism, you will be taken on an exploration of your energetic body by your guide, Naturopath, Rita Louise, Ph.D., a gifted Medical Intuitive and 25-year veteran in the Human Potential Field.

Illuminated with down-to-earth examples and case studies, Dr. Louise blends complex topics such as religion and philosophy with medicine, physics and psychology into a concise theory as to the nature of subtle energy and the disease process. It will help readers understand how disruptions to their subtle energy can be contributing to the difficulties they are experiencing in their lives. It also informs the reader how these energetic disruptions can contribute to the manifestation of illness and disease in the physical body.

In addition, readers will also discover simple steps that can be followed to help realign their bodies, minds and spirits so that they function in harmony with one another. This is when true healing begins.

To order a copies of these amazing titles visit: www.SoulHealer.com

CPSIA information can be obtained
at www.ICGtesting.com
Printed in the USA
BVHW090846030322
630461BV00008B/300